FIGHTING FOR PENNSYLVANIA

IN THE EARLY YEARS

1763 to 1783

THE STORY OF

CAPTAIN THOMAS ASKEY

AND

LIEUTENANT RICHARD GUNSALUS

OF

CUMBERLAND COUNTY

BY

Edward Leo Semler Jr.

Copyright © 2020 by Edward Leo Semler Jr.

All rights reserved by the author.

First Edition: 2020

Library of Congress Control Number: 2020910830

ISBN: 978-0-578-63867-6

Printed in the United States of America

City of Publication; Schulenburg, Texas

Cover layout by Edward Leo Semler Jr.

To the descendants of the Askey & Gunsalus family's.

TABLE OF CONTENTS

Introduction	1
French & Indian War and Pontiac's War	7
In-between Wars	23
Revolutionary War Years:	
1776	27
1777	35
1778	59
1779	83
1780	89
1781	111
1782	125

1783	131
Post War	135
Resources	139
About The Author	145

INTRODUCTION

I find ancestry and history extremely interesting, and I love to research them both. I never really start out to write a book like this one. But when I research a topic and it becomes really interesting, it just seems to grow into a writing project - which has led to this book. In fact, this is my sixth book. They all have a central theme of the military or military history. And they all involve me or my family. Two were written about my family's involvement in the Civil War. That research, and subsequent books, had surprises at every turn - to include brothers fighting against each other!

This book, however, has been my toughest to write as far as research goes. Going back 256 years is a challenge when it comes to documentation on individuals, especially detailed documentation. For the most part you had to be someone of importance to have been documented during the 1700's. If you led a major military organization or held political office, you were more likely to have details written about you or to have your letters and memoirs preserved. For the average person of the time, you were lucky to have your birth, marriage, or death recorded. Not to mention having those records survive hundreds of years.

Ink and paper were expensive and scarce, and a lot of people simply could not read or write. Documents of this era are notorious for being signed by the applicant's mark of an "X." I relied heavily on Revolutionary War pension applications writing this book and they would be dictated by the applicant to a person of the court who would actually write the application. The applicant would then sign with either their name or mark.

Luckily today we have a lot of information digitalized. Unfortunately, not necessarily categorized in an easy to research format. So going back and filtering through these centuries old documents takes time and patience. As an example, one of the best documentation of Pennsylvania's involvement in the French & Indian War is the correspondence of British Colonel Henry Bouquet. There are thousands of his correspondences which have been digitalized. And I read through each one looking for the possible mention of Thomas Askey or anyone associated with his military unit at the time. And you just can't do a search query on the document. Thomas Askey and Richard Gunsalus have about as many variations on their name as you can possibly make up. This is due in large part because the person writing it down back in those days wrote it as they heard it from the applicant. And the applicant was of course trying to remember names from over 50 years ago.

To compound things surnames had a tendency to change as people either tried to simplify them or make them sound less European. It's believed Askey was derived from Erskine, Gunsalus was spelled several different ways over the years, Semler used to be Semmler, and on my mom's side they went from Gracin to Lucas – because my great grandfather didn't think Gracin sounded American enough.

But all these hurdles are what make researching your family history fun and addictive. It's like putting that really hard jigsaw puzzle together. When you come across that one piece that ties things together it's a rush.

This book, like the others I have written, represents what I think we all wish to find in our families past; relevance and excitement!

So, when my research uncovered that my long lost relatives had forged a place in American history with their involvement in the Revolutionary War, that's relevance. Throw in that they helped establish Pennsylvania - A Pennsylvania during their lifetimes that was wild, untamed, harsh, cruel, and filled with British and native Indians wanting to kill them - that's excitement! And the formula for the makings of a book.

Thomas Askey and Richard Gunsalus, were my 6 times great grandfathers. I had always heard that we had a lineage back to the Revolutionary War on my father's side of the family because I had great aunts that belonged to the Daughters of the American Revolution (DAR). That led me later in life, when I had time to research it, to verify this and possibly become a member of the Sons of the American Revolution (SAR).

The SAR and DAR are not easy organizations to get into. You have to prove your lineage back to a person who is documented to have contributed in securing the independence of America during the Revolutionary War. And this documentation has to be verified through an intensive review process. It may sound easy, but tracking down your lineage to someone who lived over 250 years ago can be challenging.

In researching my application for the SAR I first documented my blood line back to Richard Gunsalus, verifying the family story that I had an ancestor on my father's side who had participated in the war. My 6 times great grandfather to be exact. In that process I discovered that I had a second 6 times great grandfather who also fought, Thomas Askey.

It was amazing to find I had two direct ancestors who fought. And after extensive documentation my applications were approved and I was accepted into the SAR. An honor that Thomas and Richard probably never knew they would bestow on their 6 times great grandson!

So, who were these men? Well Thomas Askey was born in Antrim, Ireland in 1727. He immigrated with his family to the colonies sometime between 1745 and 1750, eventually settling in Cumberland, Pennsylvania as a single adult. Richard Gunsalus was born in Minisink, New York in 1756. In 1773 he moved with his parents and four brothers to Cumberland, Pennsylvania. The Gunsalus family had numerous members fight during the war for New York, New Jersey, and Pennsylvania. They can trace their blood line to Spain and immigrated sometime in the early to mid-1600's.

I cannot prove Thomas and Richard ever personally knew each other, but I'm sure they had to have. Thomas was a good 30 years older than Richard and he was fighting the French & Indian and Pontiac's Wars in Pennsylvania along with George Washington when Richard was just an infant. But as the Revolutionary War heated up in the mid 1770's they lived in the same Pennsylvania County, joined the same militia, and served as officers. Then after the war they both dabbled in fighting the Indians and settled in Bald Eagle, Pennsylvania. So, it would be almost impossible for them to have never at least bumped into each other.

Their linage wouldn't align with me until Richards's great grandson John Gunsallus (they had added an "l" to their name) and Thomas's great granddaughter Pamelia Lucas met and married in 1865. John and Pamelia are buried in the Askey Cemetery in Snow Shoe, Pennsylvania along with most of their family dating back to the 1700's. And their descendants still live there today.

So let me tell you their story, a story of the founding of America. It's a factual story piecing together documents that are over 250 years old at the time I write this. I link their documented information along with events of the time to form their life story and the building of the United States of America.

This story takes place in the early 1760's towards the end of the French and Indian War. Don't worry, I'm not going to belabor the fine details of American history, I just want to refresh your memory of the dynamics at play in the mid 1700's.

At this point there was no United States of America. America was a grouping of colonies governed by the British, and therefore subjects of the British Crown and King George III. But these colonies were growing more independent. And as they pushed further west they began to fight both Indians and other European countries more as a group of Americans rather than as British subjects.

This is where our story starts, with the French & Indian War.

French & Indian War

&

Pontiac's War

The French & Indian War was fought between British and French colonies in North America. Each side had various native Indian allies as they tried to secure more territory and a stronger foothold in this new land that was basically up for grabs. The fighting lasted between 1754 and 1763.

Pennsylvania at the time didn't have the boundaries we currently know of as today. The eastern boundary was pretty much as we currently know it. But there was no western boundary, it was open and wild. The area known as Pennsylvania was split between the British claimed eastern half and the French claimed western half. Basically, the entire area outside of Philadelphia was wild, untamed, and speckled with forts. These forts provided protection for the settlers pushing westward towards current day Pittsburgh and north towards New York State. Pittsburgh was in disputed territory and marked only by Fort Duquesne, which was built by the French. It was an oasis in an extremely wild land. The French and British may have laid claim to this area, but they by no means controlled it. And not even one faction of Indians controlled it. To build a fort by no means guaranteed ownership or safety. Forts were constantly under siege by Europeans and Indians. In 1758 the British moved into western Pennsylvania and Fort Pitt was built near the site of Fort Duquesne. This was at the merging point for the Ohio, Monongahela, and Alleghany Rivers. This made this a very strategic fort because it could be supplied by three different rivers, controlled anything passing by on these rivers, and offered a perfect place to spearhead any military movement to the north, south, or west.

Travelling by waterway was probably the easiest way to get around during this time period. Roads were nonexistent west of the Allegheny Mountains and the railroad wouldn't come to America until the early 1800's. Everything was transported by boat, horse, wagon, or on someone's back.

Resupplying forts was a necessity, and lugging supplies over land was extremely dangerous, hard and time consuming. Unfortunately, the only way to resupply Fort Pitt was by land. By blazing a trail through the wilderness with an armed escort.

Pennsylvania during the French and Indian War[1]

The area of Pennsylvania that Thomas Askey served in towards the later part of the war was south of the Juniata River, west of the Susquehanna River, and around Carlisle and Shippensburg. In 1763 he was serving as an ensign with the 4th company, 2nd Battalion of the Pennsylvania Provincials.

Under the control of the British Army, the colonies had established their own defensive organizations, which were referred to as Provincials or Royal American Regiments. These organization were made up of citizen volunteers to protect the population migrating west into French and Indian Territory. The British Army had its hands full all around the world. And currently in the American colonies they were busy fighting the French. Men couldn't be spared to protect civilians moving west, so the individual colonies mustered up their own men to do it. Along with protecting their localities, they fought alongside the British, who were in overall control of the colonies.

In 1763 the problem around Carlisle and Shippensburg was with Indians as the French were operating further west. The Pennsylvania Provincials were under the control of the British Army which was commanded by Colonel Henry Bouquet, whose correspondence I quote from fairly often in this chapter. We are lucky to have these valuable documents, as they are over 256 years old. They are an amazing snapshot of what was going on in Pennsylvania during the 1750's and 60's. They detail the minute details of the period, such as how much Bouquet wanted to rent his house and land for, the proper way to make lead balls for ammunition, the daily goings on at the various forts, debriefings of civilians and military personnel who had escaped from Indian captivity. And the list goes on. I found myself spending hours just scrolling through and reading these correspondences that reveal exactly how life was 250 years ago.

They also detail how Bouquet was using the Provincials along with British and loyal Indians to push west towards their outpost at Fort Pitt. But after years of fighting the French the war was winding down, as the French drew back into present day Canada. This war was entering a new phase - Pontiac's War.

Indians not happy with the new British authority and their colonies pushing further into their land, organized under an Indian leader named Pontiac. The Indians pushed war parties into Pennsylvania at Fort Pitt, Fort Ligonier, Fort Bedford, and Fort Venango in the north western corner of the state. All of these forts held off Indian attacks except Fort Venango which was captured and burnt. War parties looking to wreak havoc were active all the way east to the Susquehanna River.

To give you an example of the Indian attacks where Thomas Askey was fighting I present to you this letter from Captain John Brady to Captain Robert Callender dated March 26th 1764 and sent from Shippensburg;

Dr Sr

I received yours & has sent ~~the~~ The War Billets.

The Captives Taken are agness Davison & one child one year old

Andrew Sims 14 Years old

Margret Stevennson 12 Years old |all on the 20th Ins^t

& Joseph Mithchel 3 d⁰

The other Damages

Adam Syms House & Barn Burn^d		
James McCalmons House	D⁰	
Will Beards House	D⁰	
Alexander Caldwells House	D⁰	the 21st Ins^t
Stephen Caldwells	D⁰	
John Boyds	D⁰	
John Stewards	D⁰	

with a Number of Horses & Cows Sheep & Hoggs Shot with Arrows at each of the above named Plantations

Walkers Barn Burned & 3 horses kill^d & one taken in the Path Valley & two horses taken away & one kill^d belonging to John Wallace in the Path Valley { the 22^d Ins^t by appearances the Indians Number that Staid after the Captives were taken away Did Not Exceed 7   I do Imagine by the Tracts I saw that the Number that Guarded the Captives away were about twelve.

Lieu^t Chambers is return^d to his Post as well as Ensign Eskey made but little Discoveries: the People reported to be slain at the Black Logg is Alive & the report Groundless.

Other Particulars Concerning Indians Seen in the cove you are acquainted with

Last Night Samuel Rippes Barn in this Town was set on fire about Eight oClock P:M: but being so Convenient to the water, was Extinguished there is a report & appears very Evident that three Indians was seen this Morning not far from Town, but who set fire to the Barn we cannot Tell but after the Party returns if we find that the Indians is yet in the Nieghbourhood, we May Easy Conjecture who Kindled the fire

I have Nothing More to inform you of,

But am

D^r S^r

Y^r ob^t Hu^l Ser^t

JN^0 BRADY

Sir/Since I seald My Letter the Indians has certainly been in the Path Valley last Sunday Morning & it appears Propable that they are not yeat gone, all the People I have is in search of them

J: B^2

Yes that ensign is Thomas Askey who this book will go into great detail. He went by Eskey and Erskine early in his life. More than likely because that is how people heard his last

13

name pronounced. They misspelled his name, which was common. You'll notice in these letters and subsequent ones that grammar was not a strong point of military men. But there is no doubt it is him. We can tell because he served with Lieutenant James Chambers who is also mentioned.

This letter reveals just how dangerous it was living in central Pennsylvania. The Indians were trying to cause as much turmoil as they could. And their only obstacle were men like Thomas Askey who had volunteered to fight them. Of course Thomas had a huge motivation - his family lived in the area. He wasn't married yet, but his father, mother and siblings lived there in Cumberland County where Shippensburg is located.

But not all Indians were hostile towards the Pennsylvania's and British. There were tribes trying to live peaceably on their land amongst the white settlers. But with frequent Indian attacks the settlers viewed all Indians as a threat. And even with an armed military presence in the area the civilian population didn't feel safe and felt a need to take matters into their own hands. Rightfully so, the Pennsylvania government looked at this vigilantly mentality and behavior as lawless and it needed to be reined in to maintain a civilized society. In a letter dated the 29th of December 1763 from Pennsylvania Governor John Penn to Thomas's boss, Colonel Armstrong he addresses this;

I am extremely surprised at the late very extraordinary Insurrections among the people in some back Counties. They have, in defiance of all Laws & Authority, assembled in Arms, marched into the Heart of Lancaster County and barbarously murdered a Number of the Indians who have peaceably resided in the Conestogo Manor for many years. And notwithstanding my Proclamation of the 22d Instant, another Party of those Rioters consisting of upwards of 100 men, came into Lancaster on Tuesday last, forceably broke open the Work House and murdered the Remainder of the Conestogo Indians who were lodged there as a place of safety.

It is absolutely necessary for the preservation of Peace & good Order in the Government, that an immediate Stop be put to such Riotous proceedings.

I do therefore hereby require You forthwith to use all the Means in your power, both as a civil and Military Officer, to discover & apprehend the Ringleaders of those Riots & their Accomplices, that they be brought to Justice; and I further strictly enjoin You to be extremely active in discouraging & suppress all such Lawless Insurrections among the People, & to give me the earliest Notice of their future Motions & evil Design.

As it is supposed, not without great Reason, that the Chief part of the Rioters live on the frontiers of Cumberland & Lancaster Counties, it cannot be doubted but, if you are diligent & strict in your Enquiries, you will soon make a Discovery of them, as they could not assemble & march in Bodies thro' the Country without being seen & known by a great Number of People.[14]

So not only was Thomas fighting the Indians and the French, he was fighting his neighbors in Cumberland County.

I don't know what inspired Thomas to join the military. It's usually done when you're a young man, when your body is strong enough to take the punishment a soldier encounters. But Thomas was pushing 40 years old, not the age one thinks about joining such an arduous profession.

When he joined, he would have been administered an oath to the King's service and in the pay of the Governor of Pennsylvania. Here is the oath he would have been given;

I, Thomas Askey, acknowledge myself to be a Protestant, and swear to be true to our Sovereign Lord King George, and to serve him honestly and faithfully within the Province of Pennsylvania, and the Provinces bordering upon it, in Defense of his Person, Crown and Dignity, against all his Majesty's Orders, and the Orders of the Governor and Commander in Chief of the said Province, and the Officers set over me by his Majesty's authority. So help me God.[6]

When Thomas joined the Pennsylvania militia in 1763, he did so during a major push by the state to stand up troops to defend against the Indian assault that was pushing east across Pennsylvania. The wages offered at the time were rank structured and in British currency. A captain was paid 10 shillings, lieutenant 5 shillings & 6 pence, ensign 4 shillings, sergeant 2 shillings, corporal 20 pence, and a private 18 pence. These wages were per day[3]. To give you some insight to how much this is, a pound, was equal to 20 shillings or 240 pence. So you were not going to make your fortune in the military, but it was a job. But not always a steady paying job.

The British had a horrible time keeping their colonial soldiers paid. If you read through the letters from the men who commanded Thomas's regiment over the years you will see a horrible track record of this, to the point of men mutinying and deserting on a regular basis. Just a few examples are from a letter on the status of the 2nd Battalion in 1758. The Colonel Commandant, James Burd stated; *"On marching from thence with a brigade of wagons under my charge, at Chamber's, about eleven miles from Shippensburg, the men mutinied, and were preparing to march, but my reasoning with them and at the same time threatening them, the most of them consenting to resume their march to Fort Loudon, where Lieut. Scott was with eight or ten months of pay."*[4] And in this letter from Lieutenant Colonel Asher Clayton, who was in charge of the 2nd Battalion in 1764, to the Governor of Pennsylvania; *"In general the troops are very uneasy about their Pay, having been Six Months in the Service, 7 by continual Scouting wore out their Clothing, without having it in their power to furnish themselves with more. Several have already deserted, & more I am afraid will follow their Example"*[5]

The men Thomas joined in the 2nd Battalion were battle seasoned, and most of the officers he would serve under fought at the pivotal battle of Fort Duquesne in 1758. Although a French victory which cost the 2nd Battalion 19 men killed and 8 wounded, the French were forced to burn and vacate the fort soon after the battle. This enabled the British to erect their own fort, Fort Pitt. This new fort on the western boundary of the colony would protect Pennsylvania for the remainder of the French & Indian War along with the subsequent Pontiac's War.

But because of the fort's precarious location out on the western frontier, it required constant resupplying. This was done in part by the 2nd Battalion. At a distance of about 190 miles, it took a good 10-11 days to make the trip from Fort Loudoun in Carlisle to Fort Pitt. And this was accomplished traversing crude paths through enemy infested terrain. As I mentioned earlier – it was extremely dangerous.

The 2nd Battalion made the trip in March of 1764, but without Thomas. He stayed back at Fort Loudoun. And in this letter from Captain Thomas Barnsley to Colonel Henry Bouquet on the 11th of March 1764 you can see that the state of the military is hanging on by a thread.

"This day I received Governour Penns Answer to your Letter, which was, that he would give his Orders for two Companys to March with the Convoy for Fort Pitt, Though he said he was in some doubt but they would Mutiny, as they had not been paid for some time nor was there any money to pay them with, and that he was much afraid they was in Want of Shoes and Blankets for the March that, tho Act by which they were inlisted Expired the first of February last and there was not other Provision made to keep them in pay; however he said I shall give directions to Col: Armstrong who is their Commander to Order them to March"[7]

They did in fact march to Fort Pitt, arriving on the 7th of April with 800 horses and without receiving a single shot from the enemy or for that matter even being discovered by them.[8]

As I have mentioned before, the low morale of these men is very evident in numerous correspondences. The leadership by the officers to keep their men focused on their mission and get them to take on dangerous situations day after day without compensation was very admirable.

For their due diligence in getting their men to answer the call to duty time after time, the officers of the Pennsylvania Provincials were rewarded with land. Meeting at Bedford in 1764 the officers requested the land to *"embody themselves in a compact settlement of some good land at some distance from the inhabited part of the Province, where by their industry they might procure a comfortable subsistence for themselves, and by their arms, union and increase, become a powerful barrier to the Province."*[9]

This land was owned by friendly Indians and the officers were asking that Thomas and Richard Penn purchase 40,000 acres from the Indians, and grant it to them. The land was purchased in 1768 and 24,000 acres given to the officers of the 2nd Battalion. Thomas received his part, lot 20, which was the upper most part of the survey and in Bald Eagle. It consisted of about 288 acres.

This was a smart move by these men. Fighting on the western frontier of the state revealed the great potential for making a living in a beautiful and bountiful part of the colony with its glorious mountains and abundant natural resources.

And Thomas was undoubtedly looking towards the future. At nearly 40 years old he decided it was time to settle down and start a family. And on the 12th of June 1764 he married Elizabeth Baker. She was the 19 year old daughter of a fellow officer, Colonel John Baker. They were married in the United Church of Christ by fellow officer, Reverend Conrad Bucher. The church had only been established a year when the ceremony took place.

The Reverend Conrad Bucher is an interesting individual. He came to the colonies as a mercenary soldier around 1755, about the same time Thomas would have immigrated. Conrad was an ensign in 1758 and more than likely with the 2nd Battalion when they fought at the Battle of Fort Duquesne. He had made lieutenant in 1760 and by 1764 he was still a lieutenant. This was an obvious disappointment to him because he wrote a letter to Colonel Bouquet in July of 1764 complaining about not being promoted and asking for his help in petitioning the Governor on his behalf. He states;

"Hon^d Sr Till now I depended, that Officers in the Pennsylvania Troops would advance by the Dates of their Commissions, but I must observe, that they do more by Favor as by Seniority, & by this Practice I have lost my Rank by a great Deal. I am now the Oldest Lieutenant on this Western Side of Susquehanna & by Right the next for a Company; but

when I have nobody, that will Speak in my Favour, I will Surely be left behind again"[10]

I'm not sure why he is out of favor with his superiors. He had been with the battalion for at least 6 years and should have made captain by then and given his own company. At this time he is in the 4th company with Thomas as the ensign and James Piper as the captain.

He was also acting as the Reverend of the United Church of Christ at this time and that may have had something to do with it. Although the church is a Protestant denomination, which was the church of King George, it may have been pulling him away from his military duties.

I think Conrad and Thomas were close and had a history with each other, and Conrad could have been the person who drew Thomas into the military. They were about the same age, immigrated about the same time, and were in the same unit. This is just an assumption I based on a lot of coincidences….

In-between Wars

As a young man of 17 years, Richard Gunsalus followed his parents in 1773 from Mamekating, New York to their new home in Derry Township, Cumberland County, Pennsylvania. This area near present day Lewistown was also known during Richards's time as the Kishocoquillas Valley. Another major geographic landmark in the area was the Juniata River. These areas are often referred to by the men of this era in their pension applications. Such as "we were men from the Kishocoguillas Valley" or "I lived on the Juniata River."

This area where Richard would spend most of his adult life is no longer in Cumberland County, but in present day Mifflin County. It's just 40 miles North West of Carlisle where Thomas spent his time during the Indian Wars. And a mere 30 miles from Bald Eagle where Thomas had been granted land and was living with his family.

Bald Eagle is where Thomas made his home after the Indian Wars. It was established as a Township in 1772 and garners its name from the local Indian chief, Bald Eagle. Thomas and Elizabeth had seven children between 1765 and 1774.

Living remotely, on the edge of what was then a vast untamed western frontier, Richard and Thomas had a daily battle with survival. Their land was intertwined with the local Indian tribes who were somewhat friendly. But still unsettled after the Indian Wars and could venture on the war path of violence at any given moment and for any given reason.

And Indians were not the only problem. In 1774 the governing British authority was becoming to controlling, and the colonist pushed back. In September and December of 1774 hostilities in New Hampshire and Massachusetts started a smoldering fire that would ignite at Lexington and Concord, Massachusetts on April 19th 1775. This was the start of what would become an 8-year war between the 13 British Colonies and Britain.

When the first engagement happened at Lexington and Concord, it was basically the colony of Massachusetts taking on the British by themselves. At this point in time the colonies were individually managed and governed, to a point, under the overall ruling British umbrella. When it was apparent that their rebellion was in need of assistance, other colonial militias came to assist. And when this happened it became apparent that their militias needed to be unified under a central military leadership system. Enter George Washington as the commanding general of the colonial military and the establishment of the Continental Army in June of 1775.

Pennsylvania units established under the Continental Army were known as units of the Pennsylvania Line. These were basically what the Provincials were back during the Indian Wars, a paid force of enlisted and commissioned men. Enlistments in the Continental Army were typically for a year or more.

Now even with the establishment of an army, the colonies still stood up and held control of their own militias. And in Pennsylvania this was a bit confusing because there were various types of militia. You had a Pennsylvania Militia and you had just plain volunteer groups, known as Associators. And these militias and associators were made up of men from individual townships and counties. They provided their own weapons, clothing, food, and usually volunteered for short periods of time – around 2 to 3 months. There was no mandatory military service, and a man was free to get involved in the military or not. These militias and associators were controlled by the county or township they joined.

Most of the duties rendered by members of the Pennsylvania Militia during the war fell into three categories. They were used to augment the operations of the Continental Army, duty on the frontier fighting Indians and Europeans in Northumberland, Northampton, Bedford and Westmoreland counties, and providing guards for supply depots located in

Lancaster, Lebanon, Reading, and at various prisoner of war camps.

1776

War had been raging for over a year in the colonies when in August of 1776 Richard joined Watt's Regiment of the Flying Camp as a sergeant. He was under the command of Colonel Frederick Watts and Captain John McElhatten. This type of unit, the Flying Camp, was established in early 1776 to fill a huge manpower gap in the Continental Army.

The Continental Army was having a recruitment problem. Not a lot of men wanted to sign up for a year or longer. The Flying Camps enlistment was for only 2-3 months, making them more attractive. Their purpose was to reinforce New Jersey while the Continental Army focused on the defense of New York. Unlike a militia or associators, men of the Flying Camps were paid like the Continental Army, even getting a bonus. Richard, as a sergeant, would be making 90 shillings a month with a 2 shilling and 6 pence bonus when serving outside of Pennsylvania. His enlistment, along with the rest of Watt's Flying Camp, would be until the 1st of December 1776.

One of the men who joined with Richard, Private James Gallaway, can give us a great insight into how being recruited happened. He states that he was out in the fields bringing in the harvest when some military men came and asked him to join. He did join, went home, made two pair of moccasins along with other preparations, and reported for duty the next day. It was that simple.

The men from the Flying Camp marched from their homes in the Kishocoquillas Valley to Carlisle where they stayed for about 6-7 days. While in Carlisle Richard was elected by his peers to 2nd Lieutenant. This was a nice pay raise of 110 shillings a month. That's how the officers were chosen, they were elected by their peers. I find this very interesting having come from a military background. You would normally think that to be an officer you had to be either well educated or well connected, or both. But these men actually picked from their peers who they had faith in leading them. I think this showed the level of democracy that they were fighting for.

From Carlisle they marched 125 miles to Philadelphia, crossing the Susquehanna River about 2 miles below Harris's Ferry and then through Lancaster. This would have taken them a good 6 days on foot. While in Philadelphia Richard was under the overall command of General Israel Putnam. After a brief stay in Philadelphia, Private Gallaway states that they *"took water"* to Trenton, New Jersey. Then through Princeton and New Brunswick and on to Amboy, New Jersey where they remained through October. At Amboy they came under the command of General Hugh Mercer.

While at Amboy the Battle of Long Island kicked off on August 27th. Private Gallaway states that in the wee hours of the morning they heard artillery in the distance. At about 4am they proceeded to march in the direction of the artillery, thinking they would be involved in the fighting. But after arriving at Paulis Hook, New Jersey, close to where the Statue of Liberty is located today, they were redirected and marched to Fort Lee, New York. All of these places are within a 25-mile radius of each other.

Soon after arriving at Fort Lee, Richard was caught up in his first engagement against the British Army. The Continental Army was reeling from defeats around New York City and General Washington, the commander of the Continental Army needed a win. Although reeling from losses he was trying to rally his army and establish a foothold. In doing this he contemplating the evacuation of Fort Washington which was strategically located on Manhattan Island. The commanding officer of the fort, Colonel Robert Magaw, felt that he could make a stand and hold the fort with the 3,000 or so soldiers stationed there, most from the 5th Pennsylvania Battalion under his command. He had success defending the fort from previous attacks, with less men, over the past few months. So, he felt confident defending it with even more reinforcements. General Washington discussed his options with his senior officers, such as General Putnam, and it was decided to let Colonel Magaw make a stand. If they could hold onto Fort Washington the Continental Army would be able to regroup and hold strategic ground.

Unknown to General Washington and Colonel Magaw was a huge intelligence breach. One of Colonel Magaw's junior officers had deserted to the British Army, taking with him plans for the fort. These plans detailed among other things, sensitive information as to where the cannons were placed. To make matters worse, sensitive letters from General Washington and his staff had also been intercepted by the British. These letters detailed how General Washington was planning troop movements and the overall low morale and lack of discipline of his men and the militias.

In early November General Washington along with most of Watt's Flying Camp were at Fort Lee overlooking the Hudson River and Fort Washington.

On the 15th of November the British decided to attack. Upon arriving, British General William Howe sent a party under the flag of truce to Fort Washington and ordered it to surrender. The British party informed Colonel Magaw that if he did not surrender, every man defending the fort would be killed. Colonel Magaw still feeling as if he held the upper hand with a strong defense, declined to surrender. So, on the 16th of November General Howe attacked the fort from various directions with over 8,000 men and several ships, known as frigates, sailing on the Hudson River.

Private Gallaway states that his company fired upon the frigates *"Roebuck and Syrian"* along with several smaller vessels. But their efforts went in vain.

By 4pm the fort had fallen into British hands. The personnel cost to the Continental Army was 59 killed, 96 wounded, and 2,837 captured. And for the British, 84 killed and 374 wounded. However, the British threat to kill everyone defending the fort did not happen and they did in fact take prisoners.

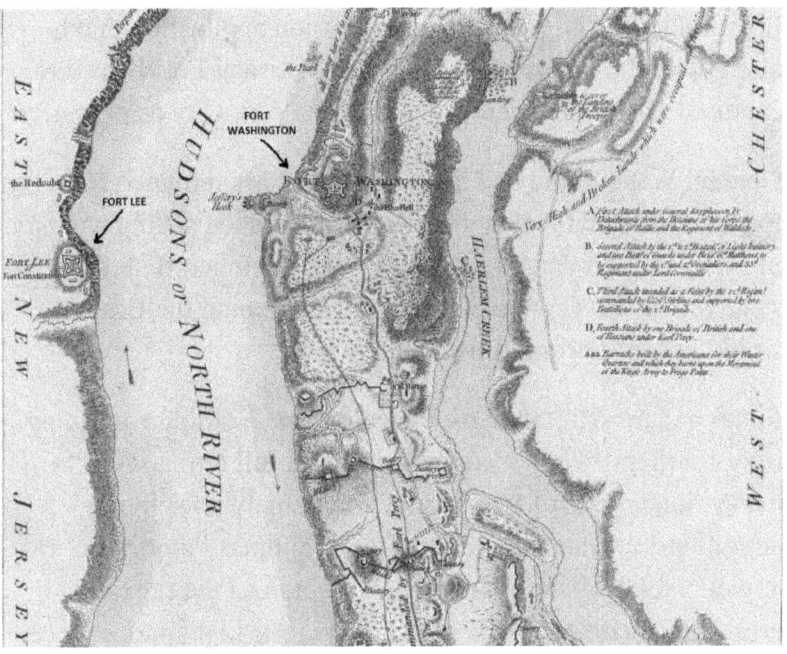

Fort Lee and Fort Washington[13]

One of those captured was Richard's Captain, John McElhattan. Since Richard wasn't captured, he had to be across the river at Fort Lee with General Washington and Private Gallaway. They watched the fight for Fort Washington unfold just over 1000 yards across the Hudson River. That must have been agonizing watching their fellow soldiers perish and be captured.

With Fort Washington lost, Fort Lee was basically useless as it could not be defended. But some of Colonel Watts' men did try and hold off the British while the bulk of General Washington's Army pulled out. One of Colonel Watts' men, Private Peter Dooey, states in his pension application that he along with the others holding off the British at Fort Lee were captured. And they were taken to New York City as prisoners.

General Washington and his remaining army retreated from the fort on the 20th of November. Richard marched south arriving in Newark, New Jersey on the 24th of November. In a full scale retreat the Continental Army were hoping the British would not follow and finish them off.

It was a depressing time for the colonists. Nothing was going right for them. They were losing and on full retreat. Add in heavy downpours of freezing rain, extremely poor living conditions, no shelter, and little food. Thomas Paine, who was a widely read author at the time was amongst the retreating army. He documented the mood at the time in a famous piece aptly titled "The American Crisis," which opened with the statement *"These are the times that try mens souls."*[11]

As the Continental Army faced one of its darkest hours and on retreat, Richard was being discharged with almost 2,000 of his fellow soldiers on the 1st of December. Their enlistment was expiring and as much as they might have wanted to stay, they needed to go home. Keeping up a homestead in the 1770's was extremely hard work. Wood needed to be cut for winter, fields needed to be plowed, planted and harvested, food needed to be hunted, and the home protected. There wasn't any time to be wasting going off to fight for 2 to 3 months, which they had just done. Now they needed to get home and help their families survive the winter.

The desire to be a sovereign country must have been a huge driving force to the men that went off to fight. Why else would they leave their homes to go risk their lives? Normally when they joined the militia or associators they stayed at home in their given towns and county. When they had to do anything for these organizations it was directly related to their own interests, like protecting their property and family. Now they were heading off to who knows where to fight and protect someone else's interests.

But after the loss of almost 2000 men and retreating, General Washington had a surprise up his sleeve. He decides to catch the British off guard at Trenton. And it's a bold move because if he fails, he could be wiped out. But the British think he has run off with his tail between his legs to winter out the war. So what better time for a surprise attack.

The Battle of Trenton unfolds in the early morning of the 26th of December. This is the famous battle in which General Washington is immortalized in a picture of him standing in a boat as he and his army cross the Delaware River into Trenton on Christmas Night. The battle was a victory for the colonists, who were desperate for a win.

In the battle James Monroe, a future President of the United States, was seriously wounded. As I was going through the thousands of Revolutionary War pension applications researching this book I came across Cornelius Dailey's application. He states that he was standing near Lieutenant Monroe when he received his wound. And that he himself had taken a "dead" ball to his left leg and a cutlass to his left shoulder. The 91-year-old veteran stated that the *"scars of which are now plainly visible."*[54]

1777

So, when Richard was discharged near Trenton, New Jersey on the 1st of December 1776 you would think he would have eagerly returned to his normal and relatively safe life back in Cumberland County. I'm sure he had marched off in August with a huge dose of patriotic pride. But he had returned with the horror and carnage of the site of war. The bone chilling feeling of not only witnessing death but of feeling it penetrate his bones night after night in the cold.

But instead, on the 1st of January 1777, he rejoined the same company and was elected as their 2nd lieutenant. This time under the command of Colonel Frederick Watts and Captain Adams as Captain McElhatten was still a British prisoner of war.

Unfortunately, Captain McElhatten would be kept a prisoner at least until the 18th of March 1780. That's the last record I can find of his status. There were frequent prisoner swaps and prisoners being paroled during the war. But even McElhatten's commanding officer at Fort Washington, Colonel Magaw, wasn't exchanged until October of 1780. There were also escapes. Private Robert Carr described in his pension application that after he was captured, he was taken to New York City. Instead of being confined like most prisoners on ships anchored in the harbor, he was assigned as a cook and waiter to his senior officers housed in private dwellings in the city. At some point, he can't remember how long, he was told by one of the officers that the cooks and waiters were going to be reduced and he was going to be sent into confinement with the other prisoners. The officer told Robert he needed to make an escape. So he gave him some money and told him to go into the market and buy food for their kitchen. While he was in the market he mingled with the locals and paid for his escape to Paulus Hook, New Jersey. And from there he made his way back to Bucks County where he lived. Once he returned back home he was eager to get back in the fight and once again joined the Bucks County militia.

The Fly Camps had been disbanded after Richard's discharge in December, so his new unit with Colonel Watt's fell under the Cumberland County Militia. They formed up in Carlisle and marched to Philadelphia. But much had happened in the short 4 weeks since Richard had been discharged. The Continental Army was on the offensive! After marching south into New Jersey General Washington headed towards Trenton, made his famous night crossing of the Delaware River, and took Trenton to the dismay of the British.

The Continental Army had also been approved to stand up an additional 16 regiments by the Continental Congress on the 27th of December. So, the defeated army that Richard had left, dragging its tail between its legs after back to back defeats, was now on the rebound of a decisive victory and a reinforced army.

From Philadelphia they marched to Levitown and then onto Morristown, where they joined General Washington's main army. In his pension request Richard mentions joining General Washington at Morristown and the death of General Hugh Mercer at the Battle of Princeton on the 3rd of January 1777. Richard states that he served under General Mercer at Fort Lee, and that Mercer had a hand in establishing the fort. I think this time period had a profound effect on Richards's memory of the war. And I think the reason is because General Washington and General Mercer both became Hero's when the Continental Army desperately needed something, or someone, to rally to. And Richard attached a memory to this pivotal moment as he was there with these men.

Up until this point in the war General Washington was not a hero. He was coming off of major losses resulting in a demoralizing retreat. The Battle of Trenton had gained back some momentum for the colonist, but it was short lived. Immediately after taking Trenton General Washington had to fall back across the Delaware River due to a lack of men to hold the town. But he clawed out respect from his men and the colonist for his daring crossing of the Delaware River in the middle of the night in horrible conditions. Not to mention taking down the feared Hessian mercenaries who held Trenton and were fighting for the British.

And to top it all off comes the victory at the Battle of Princeton only a week after Trenton on the 3rd of January 1777. During the smoke and confusion of the battle General Mercer is surrounded by the British who at the point of bayonets order him to surrender. He refuses and is bayoneted multiple times. Thinking they had actually bayoneted General Washington, they bayoneted him several more times for good measure. Mercers second in command, Colonel John Haslet is also killed by a musket ball to the head. Without leadership, and heavily outnumbered, Mercers men start a full-scale retreat. In rides General Washington on his famous white horse with reinforcements. And to add to the legendary moment, he reared back on his stallion, waved his hat, and charged into a volley of British musket flashes – not sustaining a scratch.

This is just what the Continental Army needed, multiple wins producing legendary heroes. And something like this would have had a lasting impression on Richard. Lasting enough for him to mention it 55 years later. Richard had left the fighting in 1776 with the Continental Army demoralized and in full retreat. He rejoined it in 1777 as its morale was on the rise and coming off back to back victories.

During this era of fighting army's retired to what was known as their "winter quarters" to ride out the winter months and await better weather to fight in. So in mid-January General Washington fell back to Morris Town, NJ and the British fell back to the New York City area. And as General Washington is headed to Morris Town Private Peter Dooey, who was captured at Fort Lee, states that he personally talked to him. Dooey says in his pension application that he and his fellow soldiers were treated very badly while held by the British. And that after being paroled by the British he started to make his way back home with other men from his company. While making the journey they came across General Washington and Dooey states *"I spoke to him about our situation; & he directed us to go on to Philadelphia."*[54]

Even in winter quarters the fighting didn't come to a standstill. Major battles were not fought, nor campaigns started, but instead the fighting consisted of minor skirmishes. These happened as both sides ran into each other while they were out foraging for food and supplies. Hence, the name for the fighting between January and March 1777 became known as the Forage War.

In February Richard's Company was detailed to oversee 7 prisoners, which was a major role of the militia. These prisoners were from one of the more major skirmishes during the Forage War which occurred at Spanktown.

British Lieutenant Colonel Charles Mawhood was out looking for continental foragers when he came across some herding cows. Mawhood underestimated the strength of the soldiers herding the cows and the Pennsylvania and New Jersey Line troops watching over them. Thinking he had an easy victory, Mawhood attacked. He was immediately pushed back and outflanked loosing 69 men with 7 being taken prisoner. The Continentals lost 1 man with 4 being wounded.

Richard finished out his second enlistment in April and went back to Cumberland County. His parents were probably more than happy to have the extra set of hands to help them plow the fields.

By this time Pennsylvania had made military service mandatory. This could have been a driving force in Thomas Askey writing to the Council of Safety asking for a commission in the Continental Army. At this point in his life he was 50 years old. Although just under the age limit for service, he probably could have asked for a deferral, which was common. But after his service in the French & Indian War I doubt he was going to take a back seat in this one. In his letter Thomas requests a commission as a captain anywhere Pennsylvania could use him. As a reference he states his service under Colonel John Armstrong in the Pennsylvania Provincials. Colonel Armstrong was now a Brigadier General with the Pennsylvania Militia and serving closely with the Continental Army.

The following is a great detailed description of the Pennsylvania Militia and its differences from the newly formed Pennsylvania Line serving the Continental Army;

New Pennsylvania laws *"required all white men between the ages of 18 and 53 capable of bearing arms to serve two months of militia duty on a rotating basis. Refusal to turn out for military exercises would result in a fine, the proceeds from* which were used to hire substitutes. The act provided exemptions for members of the Continental Congress, Pennsylvania's Supreme Executive Council, Supreme Court judges, masters and teachers of colleges, ministers of the Gospel, and indentured servants. Though, as a practical matter anyone could avoid serving either by filing an appeal to delay their service for a period of time or by paying a fine to hire a substitute. It should be noted, however, that a person serving as a substitute for someone else was not thereby excused from also serving in their own turn. The act called for eight battalion districts to be created in Philadelphia and in each of the eleven extant counties (such as Cumberland County). The geographical boundaries for each district were drawn so as to raise between 440 to 680 men fit for active duty as determined by information contained in the local tax rolls. A County Lieutenant holding the rank of colonel was responsible for implementing the law with the assistance of sub-lieutenants who held the rank of lieutenant colonel. Though they held military titles, these were actually civilian officers not to be confused with the military officers holding the same ranks in the Continental Army. The County Lieutenants ensured that militia units turned out for military exercises, provided the militia units with arms and equipment at the expense of the state, located substitutes for those who declined to serve, and assessed and collected the militia fines. It should be noted that these fines were not necessarily intended to be punitive. Recognizing that personal circumstances might in some cases make it inconvenient or even impossible for a particular individual to serve, the fine system was in part devised to *provide money in lieu of service in order to hire substitutes. It also provided an avenue for conscientious objectors to fulfill their legal obligation to the state without compromising their religious convictions.*

The men in each battalion elected their own field officers who carried the rank of colonel, lieutenant colonel and major and these officers were then commissioned by the state and expected to serve for three years. Within each county, the colonels drew lots for their individual rank, which was then assigned to their battalion as First Battalion, Second Battalion, Third Battalion, etc."[12]

Battalions had several companies under them and these companies were broken out into classes each time they were called up to muster. When these new *"classes were called up, each captain would deliver a notice to each man's dwelling or place of business. Under the provisions of the Militia Act, each individual summoned had the right to file an appeal asking that their service be delayed and some successfully avoided service by repeatedly filing appeals. The names of these individuals will be found on the appeal lists. The names of those who actually turned out for muster duty would then appear on company muster rolls listing the men in their new arrangement."[12]*

Contrary to the current belief that colonial America was so patriotic that everyone wanted to fight for independence, service in the military wasn't something most men wanted to get involved with. That's why there was a move from volunteer service to mandatory service. There are a host of reasons for this, one of which is the fact that the British had a large following of colonial citizens, known as Tories. These colonist supported and even fought for the British.

During the entire Revolutionary War there was a huge recruitment and desertion problem. It was common for entire units to desert. Men would either be disenfranchised by service after joining or would join to get their enlistment bounty payment and then run off with the money.

Pennsylvania tried to nip this in the bud with severe punishment for those caught. Here is an excerpt from a letter dated the 28th of May 1777 of just a few of the examples concerning the punishment handed down for desertion and/or running off with bounty money.

William Day, a soldier in Captn Smith's Comp'y 4th P.R., to receive 150 lashes on his bare back, well laid on, for Desertion, &Defrauding the United States.

Joseph Brooks, a Soldier in Captn Henderso's Com'y, 9th P.R., to receive 300 lashes on his bare back, for Desertion, and twice Defrauding the Public.

Francis Gallige, a Soldier in Captn Tolbert's Comp'y, 2nd P.R., to receive 600 lashes on his bare back, for deserting twice, reinlisting twice, Defrauding the Public, & twice Perjuring himself; - His Pay to be stop'd; to pay what Bounty he may have receiv'd from the Offier who enlisted him last, & afterwards to be return'd to the 2nd P.R.[15]

So, there wasn't a huge patriotic wave of men flocking to fight for independence. And I think this makes Richard and Thomas's involvement in the war even more extraordinary as they were in the minority of men of the time.

In late May General Washington was ready to get back to the war and moved his headquarters to Middlebrook, New Jersey which was closer to the British Army in New York. But the British Army was hastily on the move. They had hurriedly packed up on ships and had set sail, leaving only a small contingent behind - with General Washington not having a clue where they were headed. Speculation was that they were headed to South Carolina. But this wasn't the only British force General Washington had to contend with. A second British Army contingent was pushing down from Canada. And by July had captured an enormously strategic fort in the colony of New York - Fort Ticonderoga. The loss of this fort greatly depressed and concerned General Washington. To make matters worse, the British Army element that had left New York City was still unaccounted for.

By this time Thomas had received his commission as a captain and was once again under the overall command of General Armstrong, his Colonel from the French & Indian War. And on the 23rd of July Captain Thomas Askey was forming up his company of men. He was in command of the 1st Company, 1st Battalion of the Cumberland County Militia under Colonel James Dunlop.

He had added two more children to his family and now had 9 young children at home. The oldest being only 12 years old. That must have been very unsettling for him to leave his wife and young children, who were obviously vulnerable to a hostile Pennsylvania frontier.

As Captain Askey formed his men to join the fight, General Washington received intelligence that the British Army that had left New York City, was now headed up the Chesapeake Bay. They were apparently trying to make an end run on the Continental Army and attack Philadelphia! So with no other choice but to protect the Continental Capital, General Washington started to move his Continental Army south to prevent the capture of Philadelphia.

Under Colonel Dunlop Thomas marched his men from Carlisle towards the defense of Philadelphia in early August. In a letter to the Continental Congress, Cumberland County stated that they were sending out all the men they could for the endeavor to save Philadelphia. In a letter to General Armstrong from Pennsylvania's Supreme Executive Council dated 30 August 1777, they commented that;

Your Battalion of Riflemen under Colonel Dunlop, has the high approbation of Council. I am persuaded that such a Corps may be very useful. I doubt not that your zeal & patriotism will dispose you to promote the good of the service continually.[16]

This is a pretty big endorsement and vote of confidence given to the men of Cumberland County!

As they marched in defense of Philadelphia Colonel Dunlop had 8 companies consisting of 264 men in his 1st Battalion, Thomas being in charge of the 1st company.

The pension records of Private's Samuel Witherow, Samuel Quigley, John Torrence, Richard Morrow (Askey's cousin), and John Hunter shed some great light on what happened during the subsequent events. They marched to Shippensburg, Carlisle, then on to Lancaster, and finally to Marcus Hook, Pennsylvania - which is on the Delaware border. This would have been a distance of about 160 miles, taking a good 8 to 9 days to march. They state that they stayed here for a while before marching about 12 miles to Chadds Ford, Pennsylvania.

Every county was trying to throw as much as they could muster at the British attack on Philadelphia. It was of huge importance to the colonials. Not only did the Continental Congress reside there, but it was also the military headquarters of the Continental Army. So, no stone was left unturned in the defense of such an important city. Colonel Nicola from his headquarters in Philadelphia wrote to the President of the Executive Council, who was also in Philadelphia, that he was going to send every prisoner in the jail to the army. And those that couldn't be trusted on land in the army would be sent to ships in the harbor. And *"that no hands, that can be useful should be idle at this juncture."*[17.] So, everything the colonists had was thrown into the defense of Philadelphia.

As the British Army made their way up through New Jersey the Colonists planned to make their stand about 30 miles outside of Philadelphia at Chadds Ford, Pennsylvania. The ensuing battle on September 11th would draw its name from the Brandywine Creek that flowed through the battlefield. The Battle of Brandywine would go down as the biggest and longest single day battle in the war, lasting over 11 hours of continuous fighting.

As the two army's converged on each other General Armstrong along with Captain Askey were holding the far left of the Continental Army's line. They were by the creek guarding strategically important supplies. Private John Torrence says that they were not engaged in the fighting.

On the map on the following page they are in the far right corner – labeled Penn Militia Gen Arm. The British Army eventually pushed the Continental Army back. Forcing General Armstrong to fall back, as he risked being overrun.

The battle cost the British nearly 700 killed and wounded and the Continental Army and Militias nearly 1000 killed and wounded. It's hard to actually put a solid number on the colonist's figures because General Washington was notorious for falsifying his casualty numbers. This is well documented throughout the war. The reason for this was the simple fact that he didn't want high casualty numbers to reach the public. Which was understandable considering his troubles with recruitment and desertion.

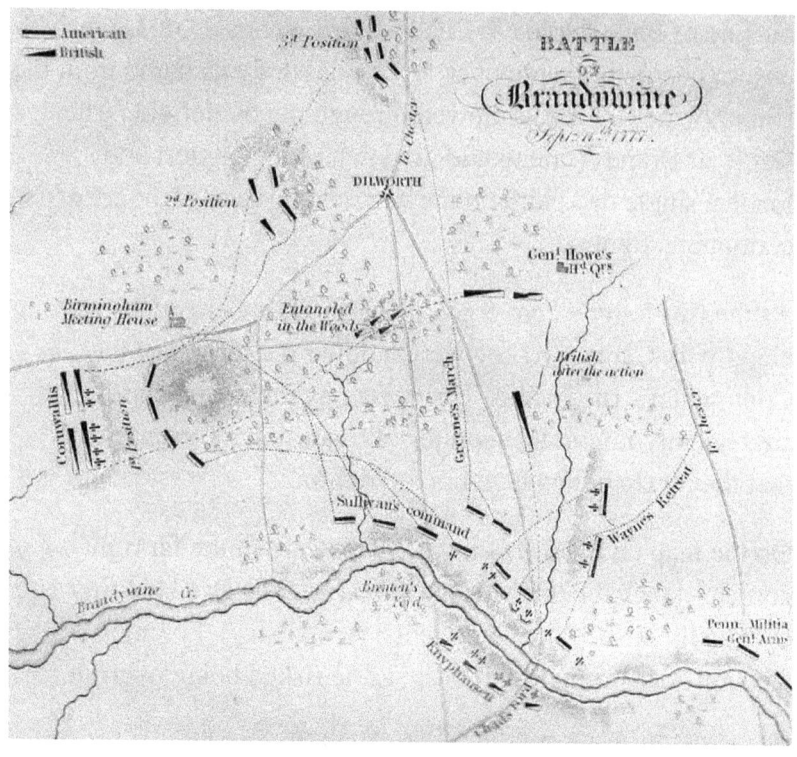

Battle of Brandywine[18]

One of the wounded at Brandywine was Private Samuel Witherow, who suffered a slight wound to his left leg. He was originally assigned to Captain Askey's company, but at Marcus Hook was detailed with several other men as riflemen under another command.

After the battle the Continental Army withdrew, ending up in Chester, Pennsylvania that night and then into Philadelphia. Moving to protect Philadelphia General Washington positioned his army between Philadelphia and the British, who were still lingering around Chadds Ford. The armies clashed again on September 16th. General Washington's Army along with members of the 3rd Battalion of the Cumberland Militia were involved in the Battle of White Horse. Unfortunately for both sides there was a downpour of rain just after the fighting began which dampened both sides gun powder. General Washington sensing his army wasn't up to a fight with the British so soon after Brandywine, pulled back. In the brief fighting before the rain Cumberland County lost Captain John Mateer.

The retreating Colonial Army could not muster a big enough force capable of stopping the British. And Philadelphia fell into British control on the 26th. It was a huge failure of the Continental Army.

Back in Cumberland County the residents were experiencing British troubles of their own. Citizens loyal to the British cause, mentioned previously as Tories, had tried to blow up the army's ammunition depot in Carlisle. Apparently several religious leaders loyal to the British and their Anglican religion, were plotting the treason. The main member being the Reverend Thomas Barton. Correspondence from the 25th of September 1777 outline the exposure of the plot before anything was actually done and the rounding up of the suspects. And I quote *"A party of Militia have taken him, and I suppose by this Time he is lodged in York Gaol. It is a Pity that men who have been employ'd in preaching th Gospel of Peace should be found engaged in such plots."*[19]

Thomas was winding down his first tour as a captain. On October 4th he participated in what would be his last action under the command of General Armstrong. Germantown, Pennsylvania was located just on the outskirts of Philadelphia. As the British went on the offensive they pushed out from Philadelphia leaving a portion of their men in the city to hold it. Seeing a chance to attack a split, and possibly weakened British Army, General Washington attacked the British in what is known as the Battle of Germantown.

In a letter from General Armstrong to the President of the Executive Council a day after the battle he gives a first-hand account of the battle;

"My destiny was against the various Corps of Jermans (Hessian Mercenaries from Germany) encamped at Mr. Vanduring's or near the Falls. Their Light Horse discovered our approach a little before sunrise; we cannonaded from the heights on each side the Wissihickon, whilst the Riflemen on opposite sides acted on the lower ground. About nine I was called to joine the General (Washington), but left a party with the Coll^s Eyers & Dunlap, & one field piece, & afterwards reinforced them, which reinforcement, by the by, however did not joine them, until after a brave resistance they were obliged to retreat, but carried off the field piece, the other I was obliged to leave in the Horrenduous hills of the Wissihickon, but ordered her on a safe rout to join Eyeres if he shou'd retreat, as was done accordingly. We preceeded to the left, and above Jermamtown some three miles, directed by a slow crossfire of Canon, until we fell into the Front of a superior

body of the Enemy, with whom we engaged about three quarters of an hour, but their grape shot a ball soon intimidated & obliged us to retreat or rather file off. Until then I thought we had a Victory, but to my great disappointment, soon found our army were gone an hour or two before, & we the last on the ground. We brought off everything but a wounded man or two – lost not quite 20 men on the whole, & hope we killed at least that number, beside diverting the Hessian Strength from the General in the morning. I have neither time nor light to add...."[21]

You can see on the following page the map of the Battle of Germantown with General Armstrong's Militia about the center of the map. Although Captain Askey's 1st Company was under the command of General Armstrong at Germantown, Private William Kelly and John Tate state in their pension applications that they were held in reserve and did not see any action.

By all accounts it appeared that the Continental Army was about to have a much needed victory. But the battlefield was blanketed in a cloud of smoke from cannon and musket fire. The advancing Continental Army units couldn't see each other through the smoke, and fired on each other by mistake. This confusion caused some units to retreat and weakened their position on the battlefield. Enabling the British to gain yet another victory.

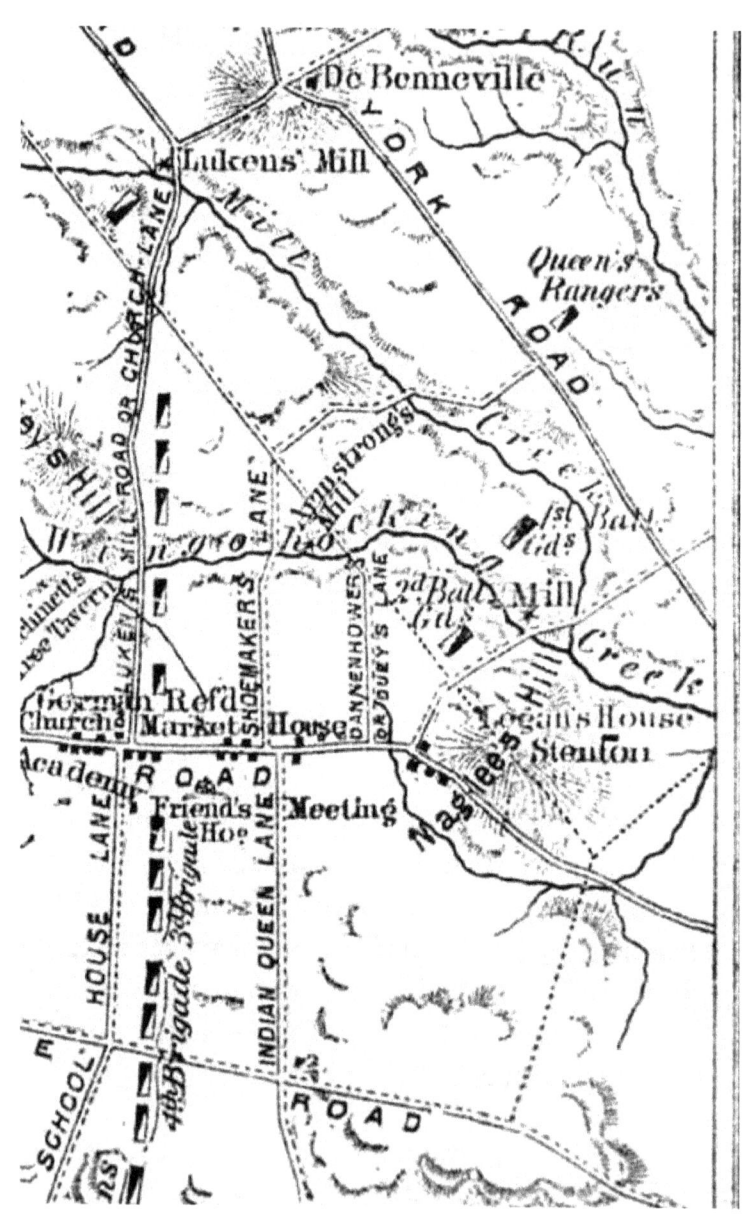

Battle of Germantown[20]

War was taking a toll on General Armstrong and his health was beginning to decline. In another letter to the President of the Executive Council on the 14th of October he states *"a late illness which has prevented my attendance at head quarters some four or five days – an obstinate Cold threatening to effect my breast & Lungs...."*[22]

Ever concerned for his men he also addresses in this letter the lack of the daily distribution of liquor that was promised to his men. He mentions that General Washington had stopped the practice as a punishment due to high rates of desertion. But General Armstrong felt they should have it back after their brave fighting at Germantown.

But with or without their liquor ration the war waged on and Captain Askey mustered his company again on the 23rd of October as General Washington once again fell back for the winter. This time to Valley Forge, Pennsylvania. Or what was referred to at the time as *"the Forge of Valley Hill"*[24] by the military. The British wintered over in New York City and Philadelphia. This gave both sides a chance to bolster their armies with reinforcements.

According to Richard Morrow, Captain Askey and his company were at Valley Forge. Richard states that the Marquis de Lafayette was at the Forge with them. And the only time Lafayette was at Valley Forge was in December 1777 and January 1778.

You would think General Washington would have a much easier task as the available man power was right in his backyard. But as I have already mentioned, the men were there, but not willing. He also had the added burden of over two thousand Pennsylvania Militia enlistments expiring the 1st of January 1778. So, in an effort to bolster the army quickly he went after the low hanging fruit, men who had already signed up and were trained, but had deserted. In his proclamation of October 1777 he issues the following;

"Whereas sundry Soldiers belonging to the armies of the said States have deserted from the same; These are to make known to all those who have so offended, and who shall return to their respective corps, or surrender themselves to the Officers appointed to receive recruits and deserters several States; or to any Continental Commissioned Officer, before the first day of January next, that they shall obtain a full pardon. And I do further declare to all such obstinate offenders as do not avail themselves of the indulgence here by offered, that they may depend, when apprehended, on being prosecuted with the utmost rigour, and suffering the punishment justly due to crimes of such enormity."[23]

So, by issuing a pardon to deserters if they came back to fight, he hoped to gain back well needed troops. And the British were not the only enemy the colony's faced. They still had the western frontier to protect from Indians.

The British had the same problem with low recruitment rates. Plus once they did round up men it was a 10 week journey sailing from England to the colonies. Not to mention they were also fighting the French, Spanish, Dutch, Kingdom of Mysore, and Maratha Empire in India. And to top it off they had to man not only an army but a navy. So man power was slim. Hence their reliance on mercenary fighters such as the German Hessians.

And as we know from the previous year of wintering over, the fighting really didn't come to a standstill. It evolved into skirmishes.

Starting with the Battle of White Marsh on the 5th of December to the Battle of Matson's Ford on the 11th of December 1777 the Pennsylvania Militia came under severe scrutiny for their conduct in battle.

Although the Battle of White Marsh was a colonial victory the Pennsylvania Militia, including men from Cumberland County, was heavily criticized for its lack of military conduct on the battlefield. In question was an incident when their commanding officer, General Irwin, was shot off his horse during a charge. His men were said to have abandoned him as he lay on the battlefield, and they retreated. As his men fled he was captured.

Second was the subsequent Battle of Matson's Ford only a few days later and just 9 miles from Valley Forge. Pennsylvania Militia - including Cumberland Militia consisting of the 1st Battalion with Captain Askey, 4th Battalion, and 5th Battalion - retreated so fast in the face of the enemy that they threw their weapons on the ground so they could run away faster.

It's unsure what part Captain Askey and his men had to play in this battle, which was commonly referred to as the Battle of Gulph Mills at the time. But in a pension application by John Barnwell he mentions being at the battle under the direct command of Captain Askey and overall command of General Potter. Both Askey and Barnwell were actually serving as substitutes at the time. Captain Askey for a Gabriel Gordon and Private Barnwell paid by the County Lieutenant.

The battle was considered a British victory with the militia losses totaling around 5 killed, 20 wounded, and 20 captured. One of the wounded being Lieutenant Thomas Blair of Cumberland Counties 5th Battalion who states in his pension application that he took a musket ball to his shoulder.

The Supreme Executive Council of Pennsylvania wrote General Armstrong in a strongly worded reprimand stating; *"The Conduct of our militia gives me real pain, Council is informed from various hands that they have behaved very infamously. The loss of our worthy General Irwin, I have been informed, was owing entirely to their base behaviour."*[25]

General Armstrong replies in two letters. In his first dated the 14th of December he hands in his resignation as the head of the Pennsylvania Militia. Although he does not mention the conduct of the militia in this lengthy letter, he does site his ill health and offers up suitable replacements for him. He obviously knew about the incidents when he wrote this first letter. But maybe he did not want to attach his resignation to the events. In his second letter dated the 16th of December he addresses the militia issue head on. Although he admits their behavior on the battlefield was infamous, he goes on to defend his soldiers;

Many, too many of the militia, are a Scandle to the military profession, a nusance in service & dead weight on the publick; yet, is it equally true, that taken as a body, they have render'd that Service that neither the State not the Army cou'd have dispenced with. They have constantly mounted guards, form'd many & distant Pickquets, perform'd many occasional pieces of labour –Patroled the Roads leading to the Enemy by day & by night & that more than their proportion – they have taken a number of prisoners, brought in deserters, suppressed Tories, prevented much intercourse betixt the disaflected & Enemy – Met and Scirmished with the Enemy as early & as often as others, and except the Battle of Brandywine, of which their Station little fell in their way, have had a proportional Share of Success, hazard, & loss of blood……Take for instance that very affair in which I agree the Cowardize of a part occasion'd the loss of Gen'l Irwin, & there we find that a very warm fire was maintain'd by others of the militia for the full space of Twenty minutes, and if we may believe report, several Waggons, some call them nine in number, were employed by the enemy on that occasion in carying off their wounded, amongst whom was Sir Jmaes Murray – two Graves were also found on the place, who or how many might be in them has not been examin'd. In the late Scirmish of Gen'l Potters Brigade, altho' they were dispersed, they must have done the Enemy some damage."[26]

The year ends with the Supreme Executive Council replying that General Potter will take over General Armstrong's command. Although relieved of his command General Armstrong still keeps himself relevant during the war and his advice is still sought after by the Supreme Executive Council.

The Continental Army is held up for the winter in Valley Forge. And the Pennsylvania Militia is cut back to bare bones manning due to the harsh winter. Although General Washington and the Supreme Executive Council have been trying to prepare their army for the winter, it is harsher than expected. General Washington issues a proclamation that everyone within seventy miles of his headquarters at Valley Forge is required to hand over half of their grain by the 1st of February 1778. And any leftover in their stock after the 1st of March will be confiscated.

1778

The New Year arrives with Captain Thomas Askey returning home from Valley Forge with his men after their enlistment expired on the 1st of January. They were part of the mass exodus of Pennsylvania Militia that General Washington and General Armstrong were so worried about losing.

General Washington had requested 2000 Pennsylvania Militia to protect his Continental Army while they wintered over at Valley Forge. General Armstrong said because of manning issues and a harsh winter he could only provide 1000.

After relieving General Armstrong, General Potter took leave over the winter lull and went home to visit his family, who were living in the unstable frontier area of Sunbury. Left in charge was Lieutenant Colonel John Lacey, who is subsequently promoted to Brigadier General on the 9th of January. It's immediately apparent in his correspondence to the Supreme Executive Council that he is extremely undermanned. His concern is acknowledged and he is promised reinforcement militia from York, Cumberland, and Northumberland County's.

By the 24th of January General Lacey is writing another discouraging letter to his superiors at the Supreme Executive Council about his situation. He is down to 70 men and his promised reinforcements have not arrived. He has his men so spread out guarding various roads that they are basically ineffective. To make matters worse while sorting through ammunition cartridges his men accidently caught the ammunition tent on fire. This caused the loss of over 7000 cartridges and 5 men were severely burnt.

By February 2nd he writes that his situation has worsened. Still no reinforcements and he has had to pull all his men off of patrols because they keep getting captured. But with all that going on I found it interesting that he still manages to end his letter requesting to know who is going to pick up the tab for his men's liquor rations!

The Supreme Executive Council in a letter back to General Lacey on the 6th of February tells him to hold his position with what he has and that his men should have been paid their liquor ration by the county they were enlisted with.

It all comes to head when General Washington writes a scathing letter to the Supreme Executive Council on the 12th of February. He lays into the under manning of the militia that is supposed to be protecting him while in winter quarters;

"The number of militia fixed upon for this purpose were one thousand, which Gen'l Armstrong promised should be regularly kept up. Upon the appointment of Gen'l Lacey, Gen'l Potter, who had been long from home, gave up the command to him. As I have not the pleasure of knowing Gen'l Lacey, I will not undertake to say whether that has been done since Gen'l Potters departure, has been owing to any want of activity in him, or whether he has not been furnished with the stipulated number of men; but this is a fact, that they have by some means or other dwindled away to nothing, and there are no guards within twenty miles of the city"[27]

He goes on to once again express his disapproval of liquor rations. Instead he has been pushing to implement a 1 shilling per day payment in lieu of liquor rations, which is approved. The Supreme Executive Council passes this news on to General Lacey. Who I am sure was not happy to have to pass this on to his men.

By the 15th of February General Lacey writes that he is now down to 60 men. But the reinforcements are starting to trickle in from York County. And I mean trickle, a dozen or so men at a time. Apparently bad weather and the fear of Indian attack had kept those form Cumberland and Northumberland from leaving home.

Once again General Washington expresses his disapproval to the Supreme Executive Council with the state of the Pennsylvania Militia in a letter dated the 23rd of February. He goes on in the letter to state that there is a huge riff between his

Continental Line soldiers and those of the Militia. But he hopes to have this worked out by the time they start to fight together again in the spring.

One of these altercations happened about the time General Washington penned his letter and is described by Private Leonard Engler of the Northampton Militia in his pension application. He states his militia *"returned as far as Allentown Lehigh County where they remained for eight days, at which place a large body of militia and regulars lay at the same time. At this place a fracas occurred by moonlight between the Regulars and Militia, which but for the timely interference of the officers would have led to serious consequences, and in the course of which Captain Foarsman of the Militia was shot by one of the regulars along the cheek and nose with Buck Shot and at the moment he fell (I) was not three rods from him. He afterwards however recovered."*[54]

Finally on the 24th of February 87 men from York County and 405 from Cumberland County report to General Lacey's camp at Doyles Town, Pennsylvania. The men from Cumberland County were men of the 5th Battalion under the command of Colonel Author Buchanan. The men from Northumberland never arrived.

It had taken the men of the Cumberland Militia almost 2 months to arrive at General Lacey's camp. A trip that should have only taken 7 to 8 days. But as previously stated, the weather and active enemy Indians had caused considerable delay. By the time they arrived their enlistment only had another 30 or so days to go. So General Lacey writes another letter stating that when the Cumberland Militia leave at the end of March he will be back down to 250 men.

When the Cumberland Militia left General Lacey around the 23th of March, they left a bad impression. General Lacey accused them of stealing horses and taking them with them on their way home. This is a huge accusation as the penalty for stealing a horse during this time was to have your ears nailed to the pillory. And it is exactly as it sounds, your ears while still attached to your head are nailed to a post for a day.

The Supreme Executive Council informed General Lacey after his complaint to them that there wasn't really much they could do without definitive proof. But they offer up some good news, he should be receiving another battalion of militia form Cumberland, Philadelphia, and York Counties.

After reading the correspondence between General Lacey and the Supreme Executive Council over the past three months, much of which I did not mention in this book as it didn't seem relevant, is that it is obvious General Lacey is not really "General" material. He doesn't really know what his responsibilities and duties are as a senior officer and is constantly complaining to the Supreme Executive Council about matters that he should be handling. And the Supreme Executive Council is constantly correcting him and guiding him on what he should be doing, things that a general should know about. For example General Lacey started to stand up his own horse brigade - and had to be reminded that he needed state approval for anything requiring funding. He also wanted to know how to handle traitors, how liquor rations were to be handled, and complained about missing horses. All things that he should be handling at his level and not bothering his civilian supervisors about.

The Supreme Executive Council was skeptical of promoting him to general at the time. But they did it more as a rushed measure to fill the gap General Potter had left and really didn't think he was going to be at home on leave for this long. When General Lacey was promoted he was only a lieutenant colonel. So, to make him a general, he skipped over the rank of colonel. Which meant he skipped over a lot of senior officers already in the field, I'm sure not making them happy. But the Supreme Executive Council felt the position required a general and they couldn't spare anyone to fill the spot. So, Lacey got the deep selection.

Meanwhile, Lieutenant Richard Gunsalus is mustered with the Cumberland County Militia under the command of Captain John McDonald in March 1778. He and his men are sent to the fortification at Frankstown, which is protecting Pennsylvania from Indians and Tories operating in the west. It's also protecting the local mines in the area which are producing lead. These lead mines were critical in the production of ammunition such as musket balls. At the same time a band of about 30 men from Frankstown decide to march west and join a force of 500 British Tories and their loyal Indians. Their plan is to attack Fort Pitt, Frankstown, and the Sinking Valley area east of Frankstown. In return for joining, the men would be given 300 acres of land of their choosing after the British have defeated the colonist. The band of men never make it to their rendezvous with the British. As they were headed that way they get attacked by Indians and their leader, John Weston is killed. Without a leader to guide them the men decide to disband and go their separate ways. Some headed back to Frankstown where they are captured and some continued on to join the British.

In his pension application Richard mentions the event stating; *"I again volunteered to go against the Indians – we were marched up the Juniata to Frankstown, we were under the command of Captain John McDonald. This applicant while at Frankstown, in company with a man named Samuel Moore, captured two of a party of Tories – the Tories were commanded by a Captain named Weston, who was killed. The Tories taken were named Hamson and Daly."*[54]

It's amazing how accurate Richards's recollection of these events in 1778 are described during his pension deposition 54 years later. I came across several letters outlining these events from General Roberdeau and John Carothers in which Richard is just about near perfect in recounting his story. Captain McDonald's men did in fact capture William Hamson and Peter Daly, and John Weston was killed - although not by the militia but - by Indians in the area. And there was an ensign in Richards Company named Moore.

It's uncertain what actually happened to the men that were captured and the ones who joined the British. In a formal interview of Richard Weston, John's brother, he states that the men were forced to go join the British. And that if they didn't they would be hung or banished to Honduras when the British won. There were trials held in Bedford after the incident and several of the men, including Richard Weston, show up in a list of men found guilty of high treason in 1784.

In April, General Washington writes the Pennsylvania Supreme Executive Council that he has received permission from Congress to stand up 5000 militiamen from Pennsylvania, Maryland, and New Jersey for the spring campaign against Britain. In response General Lacey is told he will be receiving another battalion of militia from Berks County to try and get his manning up to 1000 men.

General Armstrong now without a command, but still retained on an advisory level, writes to the Supreme Executive Council on the 13th of April. He states that General Potter has mentioned to him that he will be returning to his command at the end of the month. He also informs them that he has heard about the Cumberland Militia's horse stealing matter and has written to General Lacey about it. He relayed the Cumberland Militias side of the story - that the horses were taken as plunder from Tories and did not belong to General Lacey's command. And that he agrees that the act of taking them was not warranted, but also not punishable. General Armstrong takes a particular interest in the Cumberland Militia because he is from Cumberland. And this is where he has returned to after being relieved from command of the Pennsylvania Militia. So I am sure that when he heard the report of stolen horses that he wanted to personally address it with General Lacey.

Cumberland County isn't currently in good graces with the Supreme Executive Council either. They write them on the 24th of April on the issue concerning a lack of militia supplied to General Lacey. Going as far as calling the County "backward!" Cumberland County replies that they are sorry that Pennsylvania considers them "backward" but they have their hands full with the enemy on their frontier. They explain that they are having to send men out towards the western part of the county because there is a band of about 320 Tories operating in the area causing havoc. Plus these Tories are getting the Indians in the area all worked up and on the war path. They go on to state that they are handling the issue with their own County Militia rather than bothering the state for militia from other counties.

General Lacey receives good news from the Supreme Executive Council on the 17th of April that he is once again getting replacements. But things go bad before the men get there. He currently has roughly 330 men, almost all from Cumberland County. 165 from Colonel Abraham Smiths 8th Battalion and 147 from Colonel Frederick Watt's 7th Battalion. Colonel Frederick Watts you will remember was who Richard Gunsalus served under in 1776 and early 1777. The remaining 21 men were from Berks County. The men from Philadelphia and York County's never arrived.

On the morning of the 1st of May General Lacey was still in command, awaiting the return of General Potter. His camp was located near Neshamany Bridge on the York Road when he was unexpectedly overtaken by British troops. In this letter he details the calamity of errors, lack of military bearing from his men, and the atrocities committed by the British against his men, leading up to and during the engagement;

"My Camp was surrounded on the morning of the first Ins^t by Day Light, which lay near Crooked Billet, with a body of the Enemy, who appeared on all quarters, my scouts had neglected the proceeding Night to Patrole the Roads as they were ordered, but lay in Camp till near day, tho their orders were to leave it by two o'Clock in the morning; one of the Parties, Commanded by a Lieutenant, met the Enemy near two miles from Camp, but never gave us the alarm, he makes his excuse that he was so near them before he espied them, that he thought himself in danger of being cut to pieces by their Horse should he fire, but sends off a man to give notice that the Enemy were approaching, who did not come; on the disobedience & misconduct of this & other officers of the scouts I have to lay my misfortunes.

The alarm was so sudden I had scarcely time to mount my Horse before the Enemy was within musket shot of my Quarters. I observed that the Party in my Rear had got into Houses & behind Fences, their numbers appearing nearly to mine I did not think it advisable to attack them in that situation, especially as another Body appeared in my front to the east of the Billet, and not knowing what numbers I had to contend with, I thought it best to open my way under cover of a Wood to the left of my Camp toward Coll. Harts, which my little party moved in Columns, the Baggage following in the Rear; I had not passed before my Flanking Parties began to change shot with the Enemy, I kept moving on till I made the Wood, when a party of both foot & Horse came up the Byberry Road and attacked my right Flank, the party from the Billet fell upon my Rear, the Horse from the rear of my Camp came upon my left flank; a Body of Horse appearing in my front, we made a stand in this Woods and gave them some warm fires, which forced them to retire; their Horse suffered considerably, as they charged us and were severely repulsed; their strength gathering from all Quarters I thought it best to move on, which I did with the loss of Baggage, the Horse giving way in the Front as we advanced. We continued skirmishing for upwards of two miles, when I made a turn to the left which intirely extricated myself from them, I came into the York Road near Cross Roads, and moved slowly down towards the Billet, in hopes to take some advantage of them on that quarter where they least expected me, but I found they had retired toward the City; my People behaved well, my loss is upward of thirty Killed & Wounded, some were Butchered in a manner the most brutal savages could not equal, even while living some were thrown into Buckwheat straw, and set on fire, the Clothes were burnt on others, and scarcely one without a dozen Wounds with Bayonets & Cutlasses......You will observe in the inclosed return that fifty eight is missing, the greatest part of which I believe has run home, the rest are taken Prisoners"[28]

Known as the Battle of Crooked Billet, General Lacy was caught by the enemy with his men sleeping, not on guard duty, and he was forced to retreat in the ensuing fight. And since almost all the men under his command were Cumberland County Militia, it once again looked very badly on their character. The atrocities committed by the British on prisoners and wounded colonist were by no means the first. They had done the same at the Battle of Paoli back in September of 1777. Bayonetting wounded men over and over again as they lay wounded on the battlefield.

Those killed from the Cumberland Militia included Captain Walter Denny who died while at the head of his company, Jonathan Ross, Albert Adams, and John Neil. One of the wounded was Private Hugh Drennan from Colonel Abraham Smiths 8th Battalion. He states in his pension application that he *"received three wounds one on the head from a horsemans sword one from a musket passing through the fleshy part of his thigh and on his hand from a sword."*[54] He says he was taken prisoner and carted off to Philadelphia, was held there for six days, and made his escape. Somehow he found his company who were now in Doylestown, but was too weak to rejoin them and was sent back home.

General Potter would finally come back to take control of the Pennsylvania Militia on the 11th of May. By this time Colonel Smith was down to 102 men, Colonel Watt's, down to 110 men and General Lacey's total Militia down to 233. That's down nearly 65 men for Smith, 40 men for Watt's and 100 for Lacey in less than a month. And a fraction of the 1000 men General Washington demanded as protection.

Captain Askey would be mustered in with the 1st Battalion of the Cumberland Militia under the command of Colonel James Dunlap on the 14th of May, but they would stay and defend Cumberland County. Captain Askey, Captain Noah Abraham, their junior officers and the inhabitants of the Path Valley area, penned a letter to the Supreme Executive Council on the 18th of May about their concern for weapons, ammunition, and the safety of their families;

"That we, your Petitioners, Labor under the Greatest anxiety posseble at this present time, for our Malitia has received orders for four Classes to be in readiness to march Immediately to Camp. The Indians (or rather the tories) is Murdering our Neighbours close by us, no further off than Bedford, and what active men is of use here is Entirely Defenceless, for want of arms and ammunition. We earnestly request and beg, that the Worthy Council may take our Distressed Circumstances under their wise Consideration, and Contribute to our assistance by sending us some quantity of Rifled guns and ammunition. Likewise, to order our Malitia back against the Indians, for nothing appears to us more probable than if our men is marched to Camp our Women and Children will fall a sacrifice to Savage Cruel Barbarity. As there was a Late a Number of wicked tories Joined in a Combination, and went to Conduct the Indians Down to Murder the whigs (as they call us) here, but was Disappointed by a Supernatural Cause. Some of said party is taken, the rest

is sculking in the mountains, and thought to be the Murderers of these people Near Bedford, and their Leaders is not taken as Yet. They will bring the Indians on us if in their power. What moves us to supplicate for rifles is, because m'skets is of very little use in the woods against Indians. We hope a sensible feeling of our gloomy aspect, and the safety and security of our distressed Country and Interests, will move you to grant, with all possible speed, our Humble requests; and your petitioners shall, as in Duty bound, Ever pray, &c."[53]

Pennsylvania's Militia is really reflective of Pennsylvania as a whole at this time - in horrible shape. Roads were in terrible condition and some impassable. They were unable to get goods and supplies distributed. Civilian wagons, food, and anything else the colony could get their hands on confiscated. And in return the owners given basically worthless state vouchers instead of cash money. The British were occupying Philadelphia and pushing in from the east and Tories and Indians from the west. It really is a crucial time in the history of Pennsylvania and America.

By early June the Pennsylvania Militia is down to bare bones. General Potter is away along with General Lacey, leaving Colonel Watt's in command of the militia. Lacey laments to the Supreme Executive Council that when he returns in a few days the entire Cumberland Militia will be leaving for home, leaving him with less than 30 men!

At the same time the lieutenants that were derelict in their duties leading up to the Battle of Crooked Creek fiasco went to

General Court Martial. For their failure to post the required guards and for not sounding the alarm when encountering the enemy, they were found guilty and banished from military service. The presiding judge said *"He hopes the scandal and infamy justly inflicted on these officers may be a warning to all others, not to fall into the same Dilemma."*[29]

The British had a perfect opportunity to push into a Pennsylvania that was barely hanging on to any type of military resistance. But they don't, and decide to vacate Philadelphia and pull back and reinforce New Jersey and New York City.

By mid-June the Cumberland Militia has left General Lacey and he is down to a dozen men, most from Philadelphia County who had finally drifted in. There are no replacements sent from Cumberland County. The men they mustered up with Captain Askey in May have been busy fighting Indians and Tories in their own backyard. Even General Potter has not returned to his post to command the militia because he is busy defending his property in Sunbury, which is under constant siege. Scalping's and murders are happening daily and the Supreme Executive Council is swamped with letters pleading for support.

Then on the 3rd of July the Battle of Wyoming Valley occurred killing over 300 colonist and militia. The Wyoming Valley was a section of disputed territory in northeast Pennsylvania bordering northern Northumberland County. The area, known as Wyoming, was claimed by the colony of Connecticut and primarily guarded by their militia. British soldiers along with their loyalist Indians attacked the colonists who were settled around several forts. In the ensuing battle the British and Indian's over ran the colonists basically taking no prisoners.

The attack along with the enormously high casualties suffered by the colonists sent a wave of fear and panic throughout the area. Fort Augusta which is located on the outskirts of Sunbury, and about 65 southwest of the Wyoming battle, pleads in a letter to the Supreme Executive Council on the 12th of July;

"The Calamities so long dreaded, and of which ye have been more than once informed must fall upon this County if not assisted by Continental troops or the Militia of the neighbouring Counties, now appear with all the Horrors attendant on an Indian war; at this date the Towns of Sunbury and Northumberland are the Frontiers where a few Virtuous Inhabitants and fugitives seem determined to stand, Tho' doubtful whether To-morrow's sun will rise on them, freemen, Captives or in eternity."[30]

The Wyoming attack, along with several letters like this, cause an awakening within Congress and the colony of Pennsylvania as to the dire situation they have been neglecting; The western and northern frontiers of Pennsylvania are under siege and there is a mass exodus of people flowing out of this tinderbox that has burst into flames.

Immediately some Continental Army troops are directed from New Jersey and Philadelphia and ordered to march to Sunbury.

And a regiment on its way to Fort Pitt is redirected to assist in the fight. Militia is also ordered to muster in aid from Northumberland, Berks, Lancaster, York, and Cumberland.

The limited amount of Continental Army troops is all that could be spared from an assault General Washington was currently making on New Jersey and New York. The arrival of our allies, the French, and their powerful fleet off of New York City had consumed our army in a huge push to go on the offensive and attack.

The colonists sent 850 men to Sunbury, 450 to Easton, and 500 to Standing Stone, on the Juniata River. The 500 men sent to Standing Stone consist of Captain Askey and his company from the 1st Battalion. They are mustered up on the 14th of July and head west into the wilderness. According to Samuel Witherow, a private in Captain Askeys Company, they were providing guard duty in the Frankstown area and gaps in the Alleghany Mountains. Private James Campbell confirms this in his pension application and further states that they encountered several skirmishes with the Indians. He goes on to further state that they were mustered for 4 and a half months. This would have placed them coming back home around the first week in December.

Men from Cumberland County, consisted of Captain William Black's 7th Company from the 7th Battalion, are among those going to Sunbury. He states in his pension application that he was assigned to protect the area northeast of Sunbury and they positioned themselves about 22 miles away from Sunbury at the branch of the Fishing Creek and Susquehanna Rivers.

The British force that attacked Wyoming and thought to be descending on Sunbury never did materialize, although there were the usual small skirmishes with Indians. The British force probably didn't pursue further south because their element of surprise was lost. But their attack was effective in driving out the colonist from Wyoming and spreading fear to neighboring counties.

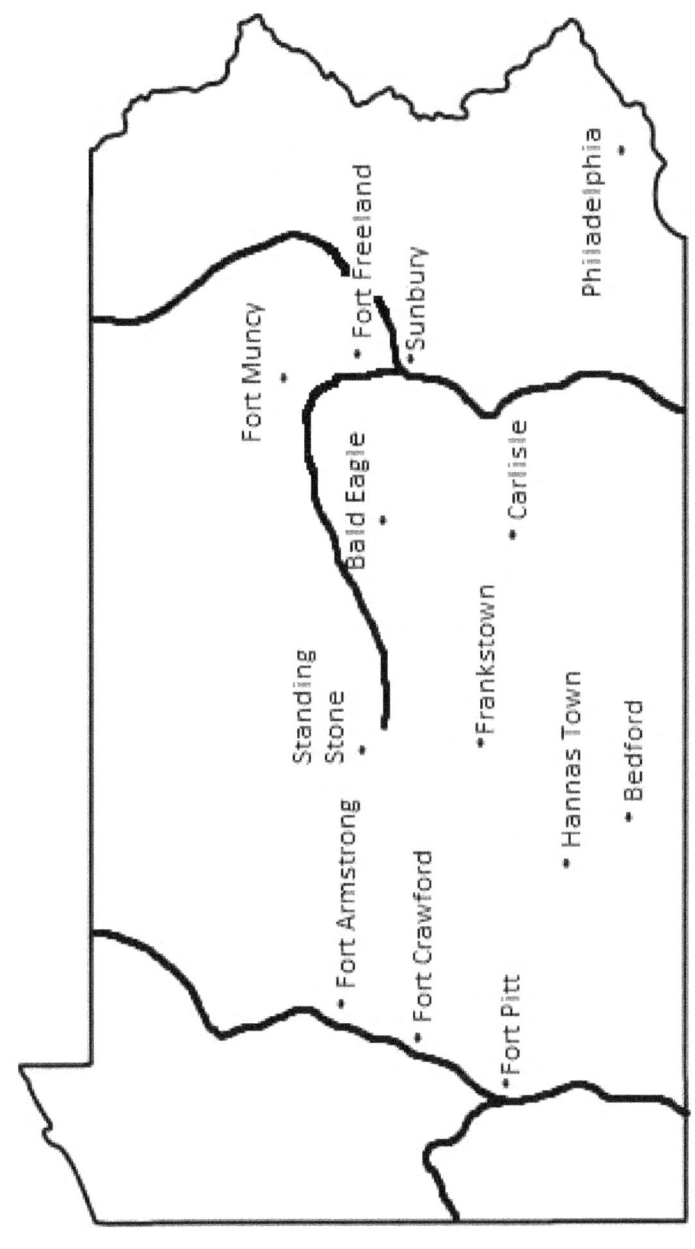

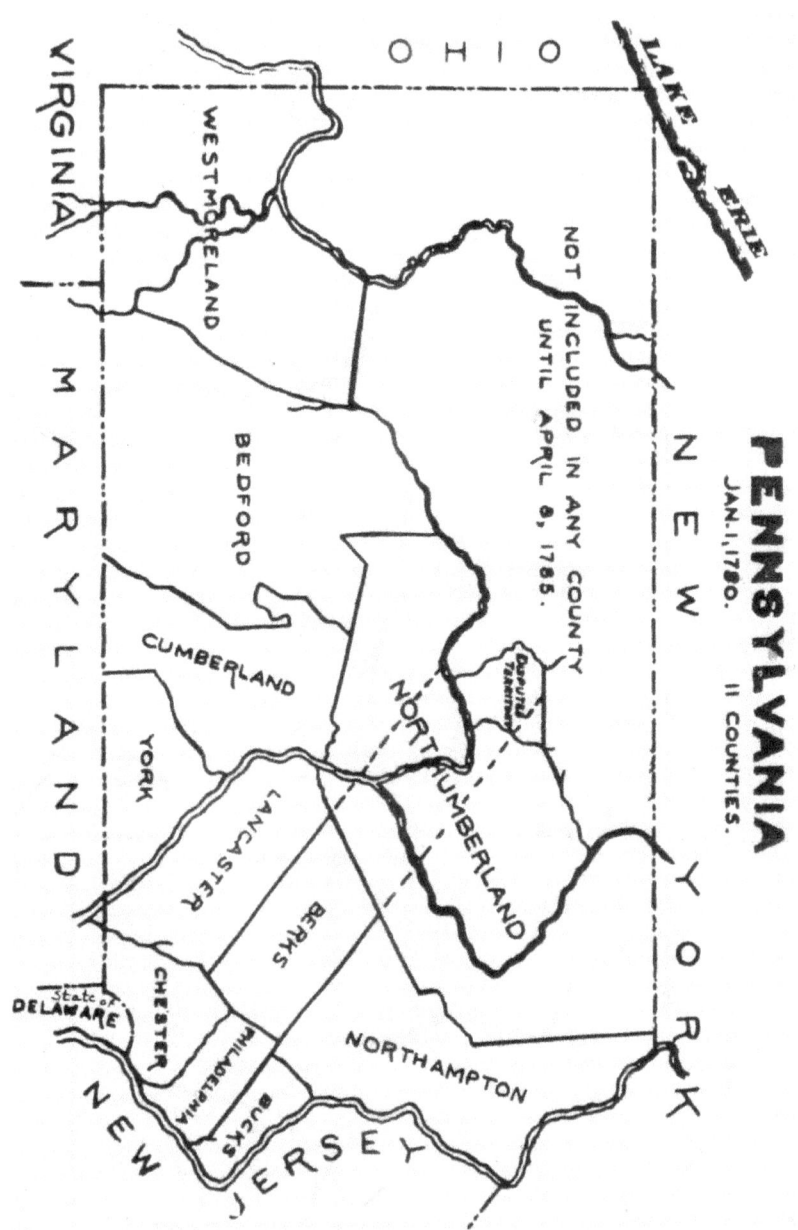

The militia as usual is having manning problems in early September. And they are urgently needed to guard prisoners. A large force of British troops pressing down from Canada were defeated at the Battle of Saratoga on the 17th of October 1777. After the victory the prisoners were kept in Massachusetts. But now due to political reasons Congress orders General Washington to move the 6,000 British and Hessian prisoners of war. It's decided that they will march them south into Virginia. As they are marching them south General Washington calls on the States Militia to escort them as they pass through Pennsylvania, as his Continental Line troops are busy fighting the rest of the British Army. And this becomes a problem for the militia.

The Pennsylvania Militia by this time is operating independently, with each County controlling its own militia, and not under any centralized command, as it previously was a few months ago under General Lacey. Counties had basically stopped suppling militia for the state's needs and were using whatever manpower they could muster to protect their own County. And if they could, they would attempt to assist their neighboring counties. This did not sit well with the Supreme Executive Council who still had statewide militia requirements. So, when the Supreme Executive Council charged the counties in which the prisoners were passing through to provide their militia to guard them, the county's baulked at the request. They stated;

"That the Invasion of the State the last Year by which the militia were Subject to constant & very severe service makes it

an Act of Justice to give them all possible indulgences, as their agriculture & necessary Attention to their Families was very much interrupted by their military Duty"[32]

It's not that the county's really had a choice. A few months prior to this Major General Benedict Arnold requested that the militia be raised to guard Philadelphia so the Continental Line troops there could be redirected to the fight with General Washington in New Jersey and New York. They called 400 men to militia duty and only 80 showed up.

A problem on the extreme other side of the state was the amount of time it took to get militia assistance out to far away settlements like Fort Pitt. Counties on the western frontier relied on other counties to assist them since they were so sparsely populated. General J. McIntosh writes the Supreme Executive Council on the 29th of December 1778;

"I must observe to you, that all the Militia I had were from the State of Virginia & none from Pennsylvania, nor would they be of any Service if they were willing, & had Joined me, as your present Militia Law, I understand, allows, or, which comes to the same thing, does not oblige them to serve above two months, one half of which will commonly be taken up in collecting them together & the other half with Incumbrances, Disappointments, &c."[33]

As 1778 comes to a close the British are trying to maintain their footing in New York City, have an offensive in Virginia and the Carolinas, and have to deal with the French who have entered the war on the side of the Colonists. This means they have to protect assets in Europe, India, and the Caribbean along with those in the Americas.

Pennsylvania's Militia is also stretched to a breaking point. The limited amount of men wanting to fight are primarily joining the Continental Line because they pay more and offer bounties. Whatever men are left get picked up by the County Militia's, which are not many. Three years into a seemingly endless war has left almost every man of fighting age hesitant to join the military or only fall in when their mandatory militia time comes. The priority of most men is the protection of their own family and homestead.

1779

The problem of British soldiers, Tories, and Indians operating on the western side of Pennsylvania was a major issue entering 1779. By April more militia were ordered west to Bedford and Westmorland Counties from York, Lancaster, and Cumberland Counties. Knowing that mustering and sending men was an issue with the counties, the Supreme Executive Council reassured the counties that the militia sent out would only have to stay in service until Continental Line troops and Rangers could be sent to relieve them.

Rangers were a new type of militia, or soldier, which would be enlisted for 9-month periods. Rangers were to be a happy medium between the militia, which usually signed up for 2 months, and Continental Line troops, that signed up for 1 to 3 years. The Pennsylvania Rangers would operate exclusively on the frontier areas of Pennsylvania. They would be in service long enough to muster, march to where they were needed, and be on scene long enough to have an impact on the problem at hand. And they would have a more relaxed operating tempo that would be appealing to frontier men, who were shunning structured military service.

While nearly 900 Rangers were being recruited and formed for service in May, General Washington recommends that the entire effort of fighting the enemy on the Pennsylvania frontier be led by General Potter. His experience at Sunbury and the Wyoming area of Pennsylvania had impressed General Washington, and Washington recommends him as his choice to the Supreme Executive Council.

I found that endorsement for General Potter pretty interesting since he never really filled his position as the commander of the Pennsylvania Militia and spent most of his time home in Sunbury while General Lacey ran things. But he did serve under Washington at the Battles of Trenton, Princeton, Brandywine, and Germantown. So I'm sure that allegiance was rewarded with this endorsement.

But men still could not be mustered into the militia to fulfill any type of obligation. Lancaster County writes in what is a common tone among the counties;

"I have called upon the class in rotation of four Battallions & from the returns made me find only thirteen privates willing to march……The Militia of this County have been much harassed with the sudden marching of Prisoners, furnishing Sundry Guards over Continental Stores in different parts of this Country, as well as facing the common enemie."[34]

By May 24th Captain Thomas Askey once again mustered his 1st company of men to head off to Hannas Town to assist Westmorland County. They would be serving under Colonel James Dunlap's 1st Battalion of Cumberland Militia. But it wasn't easy getting them together. Cumberland County wrote to the Supreme Executive Council;

"The Militia Ordered by council to Hannaha Town in westmoorland county, have with much Difficulty and Delay, proceeding from various causes, been sent forward. – the ravages of the enemy upon the Susquehanna river, bordering on the fronteers of this county, have in a great measure prevented the militia of three battalions now called to serve at hannahs Town, from turning out. They aledging it very hard for them to be sent away so far and leave their own fronteers and families exposed to the enemy, Who are murdering and burning every week within twenty or thirty miles – They say they are willing to serve upon their own fronteers, but to be Taken so far away and leave their own families in such a Dangerous Situation, they will not submit to, let the consequences be as it will."[35]

And Captain Askey and his men would be traveling far from home. Hannas Town was just east of Fort Pitt. That's 170 miles from Carlisle and a good 8-9 day march over rugged and enemy infested country.

Captain Askey and his men finished their tour without incident and returned to Cumberland County. They had just returned home when they heard of fighting going on at Sunbury, not far from their homes. A band of over 300 British troops and loyal Indians had once again been raining terror in the area. Their most recent attack occurred on the 28th of July at Fort Freeland just a little north of Sunbury. There they killed over 16 soldiers including two captains and obtained the surrender of the remaining men and women who had sheltered in the fort. This hit close to home and the men of Cumberland County scrambled to send men to help in the fight. They muster their 3rd Battalion under the command of Lieutenant Colonel Samuel Irvine and marched to Fort Freeland in response to the attack. Also joining the fight was Captain William Black who I have mentioned earlier. He wanted to help so badly that he reduced himself down to a private, yes private, so he could serve under his fellow 7th Battalion friend, Captain James Power. Power was able to quickly gather his 2nd Company men and they headed off towards Fort Freeland. Once they arrived at the fort they chased the enemy north until they reached Fort Wallis, and gave up pursuit.

Just an interesting note here on Fort Wallis. In the official correspondence it's stated that the Indians were chased to Fort Wallis. Fort Wallis was just a stone house built in 1769 and the actual fort there was Fort Muncy, which was built in 1778. In several pension applications I read by Abraham Dewitt and Philip Dinturff they claim that they built Fort Muncy in 1778 and the Indians, according to Dinturff, were chased until Fort Muncy.

The order goes out again for militia in the area to muster to their neighbor's defense in Sunbury. Fresh off of their last mustering Captain Askey's men are called together on the 1st of September. They are under the command of Captain Alex Peeples and they march to Sunbury.

Upon the arrival of Captain Askey and his men things quiet down around Sunbury. The enemy that attacked Fort Freeland retreated back into northern Pennsylvania. There was a Continental Line regiment and Rangers operating in the area so the Cumberland Militia was playing a supporting guard role and was not on the offensive.

Meanwhile the main northern arm of the Continental Army was still engaged in the New Jersey and New York City area. The British were heavily dug in making an attack impossible. So a sort of stalemate occurred as each side tried to obtain intelligence on what the other side was up to and where their weakness was. Most of the intelligence obtained by the Colonists was derived from enemy deserters. Some of the deserters were British and Hessian soldiers and some were Colonists forced to fight for the British. Even though desertion was punishable by death, the deserters felt it was worth the risk to get away from miserable conditions being imposed by the British.

As winter set in the armies once again started to settle into winter quarters. And Captain Askey and his men returned home from Sunbury in November.

As the militia settled down for the winter, Cumberland County was in need of a County Lieutenant. This was a position of authority over military matters for the county. Usually a civilian position the occupant held the military rank of colonel. One of the jobs was to oversee all militia matters and report to the Commander and Chief of the State Militia - the Supreme Executive Council. General Armstrong recommend to the Supreme Executive Council that Captain Askeys boss, Colonel Dunlap, be selected for the position but he declines it.

General Washington takes the time in early December to commence a military court martial in Morris Town for Major General Benedict Arnold. He is accused of misconduct with enemy sympathizers while overseeing the occupation of Philadelphia after the British vacated it, a position he still held. In a nutshell some of the wealthy residents of Philadelphia were accused of siding with the British during their occupation. Once the Colonists took Philadelphia back, General Arnold was accused of being too cozy with this wealthy lot, even marrying one of their daughters.

Things weren't looking good for General Arnold. But nor were they looking good for General Washington and his Continental Army. He pens a letter to the Supreme Executive Council on the 16th of December that he has run out of food for his army at his winter quarters. He states that the conditions of his men are the worst he has encountered and if he doesn't receive relief his army will have to disband within two weeks. He backs that letter up with one to Congress on the 5th of January 1780 stating that his men have gone without meat for 5 days and desertion is on the rise.

1780

The New Year starts off looking very bleak for the colonists - the Continental Army in Morris Town is in dire need of food and clothes and the Pennsylvania Militia is undermanned and unable to recruit men to fight.

Making things worse for the Continental Army is extremely bad weather. Severe snow storms have left roads untravelable, making it impossible to get food and supplies to the army. Some things have been trickling in, but not enough.

The Continental Army is still fighting the war on two fronts, the northern and southern campaigns. Pennsylvania is still heavily supporting the northern campaign with Continental Line troops. And the militia is busy with frontier duties, protecting their own County's, and assisting the Continental Army.

And if fighting multiple enemies wasn't enough, infighting breaks out within the colonies. Things get a little heated between the Continental Line troops operating in the western part of the state and the Rangers. The Commander of the western frontier forces, which was the 8th Pennsylvania Line Regiment, was Colonel Danial Brodhead. He was getting ready to make a push into the northwestern part of Pennsylvania and then into New York. He had recruited a bunch of men assigned to the Rangers to join him, but there was a problem. The men who enlisted with Colonel Brodhead still had several months left on their enlistment contract with the Rangers.

He ordered the men he had recruited to fall into his command at Fort Pitt. The commander of the 1st Ranger Company Captain Joseph Erwin and commander of the 2nd Ranger Company Captain Campbell both declined to let these men leave until their enlistments with them were up.

These men were already stationed at Fort Armstrong and Fort Crawford on the Alleghany River north of Fort Pitt. And these areas were hot beds of enemy and Indian activity. So, it's understandable that the Rangers didn't want to lose any men defending these areas. Neither did the County Lieutenant of Westmorland, Colonel Archibald Lochry. He backed the Rangers and refused to let them go. And additionally he ordered them further back into Westmorland County to Hannas Town and Fort Crawford.

This infuriated Colonel Brodhead who had Captain Erwin and Campbell arrested! Well arrested in absentia, he sent them notices stating that they were under arrest and they needed to report to Fort Pitt for trial. Neither went to Fort Pitt and they obliged Colonel Lochry.

Meanwhile back on the eastern side of Pennsylvania General Arnolds Court Martial was coming to an end in late January. A letter from the Supreme Executive Council to Congress begged for leniency in the matter as they felt General Arnold was an upstanding General. He did in fact lead the Colonists to several major victories early on in the war. So it's understandable that on the 26th of January he was acquitted of all but a few minor charges.

Trying to ease the tensions with its own military on the western frontier the Supreme Executive Council writes to Colonel Brodhead on the 14th of February in a very delicate and diplomatic fashion that it sides with the Rangers. They dance around and bolster Colonel Brodhead's pride by telling him that his *"Zeal to inlist them we highly commend"*[36] and we *"retain a great personal Respect to your Character & Service."*[36] Because they are of course walking a tight rope of trying to keep both, extremely essential front line, military units happy.

As March ends the Continental Army in Morris Town is barely hanging on. Generals are writing General Washington that their men are deserting and going back home for lack of food, clothing, and shelter. Foraging has exhausted all means of local provisions. The situation is so dire that Congress issues a proclamation that April 26th will be a *"day of fasting. Humiliation and prayer, that we may, with one heart and voice, implore the Sovereign Lord of heaven and earth to remember mercy in his Judgments."*[37] It was reminiscent of Jonah and his proclamation to the people of Ninevah to repent or be destroyed.

Every day the Supreme Executive Council is receiving letters of urgent requests from General Washington at Morris Town, General Brodhead at Fort Pitt, and a host of other people - military and civilian. The situation looks as worst as it has ever been for Pennsylvania and the budding "Americans."

And a letter from William Maclay in Sunbury on the 2nd of April drives home a sentiment that is really keeping them going. He of course starts his letter off with how dire things are in Sunbury, but ends it with *"Thus is a strange divided Quarter – Whig Tory, Yankey, Pennamite Dutch Irish and English Influence are strangely blended. I must confess I begin to be national too and most sincerely believe every publick Interest of America, will be safer in the Hands of Americans, than with any other."*[38]

In early April the Supreme Executive Council establishes a special procuring and foraging law to try and get food and hay to the military. Their procuring officers immediately start the procurement of these items and its distribution to the army at various stocking points. As an example, General Washington requested that Carlisle receive 800 barrels of flour, 4,000 gallons of rum, 80 tons of hay, and 4,000 bushels of corn. But there was a downside to feeding the army - the state goes 20 million dollars in debt with credit certificates to vendors.

Pennsylvania touts the certificates are better than gold or silver as they have a 5% interest rate and good for 10 years. But a majority of venders need cash immediately to pay their workers and to buy material. So, some procurement officials front their own money to do this. People like William Henry of Lancaster shells out $200,000 of his own money to keep a shoe factory in production so it can fill an order making shoes for the army. He of course seeks reimbursement. But selfless acts like this keep companies afloat and goods being produced for the cause of defeating the enemy and independence.

They also needed to get the Pennsylvania Line units formed back together after the lull over the winter. Most units had lost men due to desertion. To entice these men back, attract new recruits, and have current soldiers re-enlist, the Supreme Executive Council revised incentives for joining the Line. Officers were given clothing allowances and all soldiers given rum, sugar, tea, coffee, tobacco, and soap allowances. Officers that were receiving ½ pay pensions for 7 years after their service were extended to life. This pension was also granted to those disabled in battle. Land grants were also given in appreciation for service. A major general would receive 2000 acres, brigadier general 1500, colonel 1000, lieutenant colonel 750, surgeon 600, major, 600, chaplain 600, captain 500, lieutenant 400, ensign 300, sergeant 250, and a private 200 acres. This land was tax free to the soldier and would be passed on to the heirs of any soldier killed in action.

But even with these incentives the Pennsylvania Line is a calamity of errors. General Irvine, the commander of the 1st

Commissary Brigade, writes to the Supreme Executive Council on the 2nd of May that he has mass desertion. He also laments that officers who have served their enlistment and are no longer in service are writing certificates for men stating that their enlistments are fulfilled, or over, when in fact they are not.

The Supreme Executive Council writes to Chief Justice Thomas McKean;

"The Desertion of the Army is very alarming & the Officers complain that when they go into the Country they get little Assistance from the court officers."[39]

County Lieutenants on the western frontier are continuing their demand for militia from other counties to come and protect them. But the militia is also a continuing problem. The promised relief from the Continental Army is never truly fulfilled as they are stretched to the limit fighting on two fronts. And the Rangers are slow to form. So it is brought up by one of the County Lieutenants to just form up small raiding parties like the enemy. The concept was that even if the entire Continental Army was available, it couldn't cover the area required to be protected. So why not just form up small bands of fighting men to go out on the offensive and attack the enemy as they attack them, in small bands. This idea is embraced by the Supreme Executive Council and they push this concept to the Counties. To sweeten the deal they also throw in a bounty on enemy scalps. This was basically a call for mercenaries to operate against the enemy. A prime example of this is in a letter to Captain Nathan Boys on the 2nd of May concerning enemy boats operating against colonial boats on the Delaware River;

"You are with all Expedition to engage Volunteers not exceeding 30 exclusive of Officers, & proceed down the River Delaware in Pursuit of several Boats manned by Refugees from New York whom you are to treat as the most inveterate Enemies of these States. As an Encouragement to these Volunteers they shall receive Provisions, Arms & Ammunition, from this State for the Arms & Ammunition the Officers are to be accountable on their Return, and for every Prisoner taken or Man necessarily killed you shall be entitled to receive 1000 Dollars, the same to be divided among the Officers & Men in the following Proportions:

Commanding Officer	*4 shares*
2nd, in command	*3 shares*
3d, in command	*2 shares*
& every private, one share"[40]	

The Delaware River is the dividing line between Pennsylvania and New Jersey. As it runs north out of Philadelphia it flows through a volatile wilderness filled with Tories and Indians. And in this area Richard Gunsalus's cousin, Manuel Gunsalus, has his house attacked in early April by four Indians. His house is in Northampton County on the banks of the Pennsylvania side of the Delaware River. During the attack the Indians take him and another man, a servant, by the name of

John Dayley as prisoners. Surprisingly his wife and kids were at home but were not harmed. Neighbors rallied and went out after the Indians, but Manuel and John were not found. After attacking Manuel's house, the Indians attack several other houses a few days later. Local military men under the command of Captain Westbrook ran them off with casualties on both sides. The County Lieutenant stated after the attacks to the Supreme Executive Council that;

"the people are Determined to Evacuate the Country, as there appears no respect to Relief by the militia."[41]

One of the men serving with Captain Westbrook was Samuel Gunsalus, who was also a cousin of Richard, and possibly Manuel's brother. He was working as a scout in the area for Captain Westbrook and probably took part in the endeavor to try and rescue his brother. Samuel went on to live a long life, applying for a Revolutionary War pension in 1832 at the age of 75 years old. His service besides scouting was with the New Jersey Militia from 1776 to 1778 where he participated in several battles.

It's not known what happened to Manuel or John Dayley. More than likely they were killed by their Indian captors.

On the 10th of May Captain Askey and his men are once again mustered along with another battalion. The Cumberland County Militia has gone through a restructuring and he is now in command of the 8th company of the 6th Battalion. One battalion is sent to Sunbury in Northumberland and the other to Bedford in Westmorland. It's unclear which location Captain

Askey was sent. But there are serious problems in both Northumberland and Westmorland Counties. In Northumberland the town of Sunbury has been protected by the German Regiment, a unit of the Pennsylvania Line. Its Commanding Officer, Lieutenant Ludowick Weltner, wrote on May the 6th that he only had six more days of provisions. And if he didn't get provisions he would abandon his post. And in Westmorland, a captain and 11 of his men were killed by Indians and there was an urgent need for soldiers.

In a letter on May 30th from Major Robert Cluggage to Colonel John Piper he discusses men from the Cumberland Militia guarding the gaps in the Allegheny Mountains near Frankstown. These were more than likely the men that went to Westmoreland County. And this was more than likely where Captain Askey ended up. I say this because he and Richard Gunsalus seemed to spend a lot of their militia time in the Frankstown area.

At the same time, General Washington is trying to establish enough of an army to go on the offensive. He says he needs 20,000 men in the Continental Army, but he can only raise 9,000 and only has 3,500 in Morris Town. And he will not rely on the militia to help. He writes on May 25th;

"Militia cannot have the necessary habits nor the consistency either for an assault or a siege. In employing those essentially, we should run a risk of being abandoned in the most critical moments."[42]

I'm sure the mass exodus of militia during his campaigns in December of 1776 and January of 1778 were still fresh in his mind. Not to mention the fact that his regulars and the militia keep bumping heads.

The Pennsylvania Militia at this point in time was still mandatory service, but in actuality few served. And those that did so, like Thomas Askey and Richard Gunsalus did so many times as substitutes for others. The militia law was written in order for you to have a paid substitute fill in for you, and nearly half of the militia were substitutes. Men made a living doing their mandatory service time and then getting paid to do someone else's.

As General Washington was trying to gather his army to go on the offensive, the British had the same idea. Receiving intelligence that the Continental Army in Morris Town was weak, plagued with desertion, and low morale - the British decide to go on the offensive. Leaving their positions in New York and New Jersey they started to move towards Morris Town to attack. New Jersey Militia and Line troops engaged the superior British advancement on June 7th at the Battle of Connecticut Farms. The British were surprised by the strong resistance they met so early in their campaign. This along with the fact that General Washington was now onto their plan caused them to pull back to their original position.

General Washington takes a minute to praise the militia that more than likely saved his army. He writes to Congress on June the 11th;

"The Militia of this State have run to arms, and behaved with an ardor and spirit of which there are few examples."[43]

If it wasn't for this valiant stand by the militia his army could have been wiped out. Not only did he have all the problems previously mentioned, but he also didn't have enough horses to move his artillery and supplies. After this initial attack he had time to round up and confiscate horses to get his army moving. Which was wise because the British tried once again to advance on June 23rd.

And General Washington was concerned. He writes on the day of the attack to Congress;

"But we do not know what may be the ultimate design of the enemy – all we know is that they are very strong and that we are very weak."[44]

But, once again the British are met by surprising resistance from a small number of Continental Line soldiers and New Jersey Militia at the Battle of Springfield. The stiff resistance slowed the British advance. Surprised once again by the enemy they retired back to their original position.

By mid-July General Washington's army was still struggling to form up into a fighting force and he laments to Pennsylvania that they have only sent 1,000 men so far. He fears that the fighting season is slipping away from him. He also worries that our allies, the French, are ready to fight but we aren't.

Captain Askey and his men return home from their two month militia duty. The frontier Counties are spared from sending men to aid General Washington. They are being sourced from Philadelphia, Berks, Lancaster, Bucks, and Chester Counties.

But if getting men to join the militia was hard before, the bar was raised again with the extremely lucrative new deal Pennsylvania was now offing men to join the Pennsylvania Line;

"Being anxious to promote voluntary Inlistment & fill up the Line of Pennsylvania we have concluded to attempt inlisted Recruits for the War for a Bounty in Land & Specie, viz. 200 Acres of good Land & 3 half Johannes for every able bodied Recruit, free from the Rupture Lameness or other Disorder, not more than 45 nor less than 18 Years of Age.

No Deserter from the British Army or Navy or Prisoner of War Apprentices or intended Servants to be admitted to enter.

You will also avoid inlisting Sailors & Foreigners, and more especially Frenchmen on any pretext.

Besides the above you are authorized to promise a Suit of Cloaths yearly & Blanket, a Pint of Rum a week with Tobacco, Soap & sundry other Necesaries with Pay & Rations as the Continental Troops - They are to serve under the immediate Command of Gen. Washington.

As an Encouragement to the Sergeant he will have a Spanish Dollar or 60 Dollars Continental for every Recruit he inlists who passes muster. And the Officer under whose Direction the Inlistment is made will be entitled to 200 Dollars Continental Money for his care & Trouble.

You will be particularly careful not to Suffer any unfair Practices of catching Persons by putting Money in their Pockets, or such like Acts, but inlist them fairly & openly.

Every Recruit, Care being taken that he is quite Sober."[45]

It's probably not lost on the men that are being sought for recruitment that Pennsylvania can't honor what they are offering. The 200 acres, probably - but it's on the frontier where you're going to have to fight off Indians harder than the British. The 3 half Johannes, which is a gold coin, is doubtful. Pennsylvania is essentially broke and can't even afford to clothe and provision the existing army, not alone the new recruits. Letters are constantly pouring into the Supreme Executive Council about having to go without. Lieutenant Weltner in Sunbury states he has been without Rum for over a year and is embarrassed that he can't offer anyone who visits him a drink. Colonel Henry Haller who is the commander of the gaol (jail) in Reading says he will have to let his prisoners go if he isn't provided provisions. And as I stated earlier the Continental Army just barely squeaked by the past winter for lack of provisions and clothes. So, I think everyone knew the offer the Continental Army was making couldn't be fulfilled.

I pounder why Thomas Askey and Richard Gunsalus never joined the Pennsylvania Line. With as much service as they had provided to the militia they could have done an enlistment with the Line. I would think the benefits, although a bit of a farce, were still better than the militia. Thomas would have probably been too old to join the Line at 53 years old in 1780. But as a career officer I'm sure he could have received a waiver. Richard was 24, well within the age standard, and was single, so he was a prime candidate. My only conclusion is that they were too worried about leaving their families alone on the open frontier that was constantly being threatened by Indians and the British. Even without the British threat the Indians were still there and still fighting for their land and sovereignty. Doing their militia commitment of 2 months was enough to serve and protect their community. And filling in as substitutes put extra money in their pockets. Not as lucrative as joining the Line, but they probably knew they were not going to see the incentives offered. And they didn't want the commitment of serving out a long enlistment – that could possibly last the duration of the war, which had already been dragging on now for over 5 years.

The extravagant offer to join the Pennsylvania Line didn't draw as many recruits as needed, so the Supreme Executive Council sends a directive out to the frontier Countys, including Cumberland. They are told that they need to call out three classes of militia to be ready at a moment's notice to march off with General Washington. Understanding that Cumberland County is still dealing with Tories and Indians, they explain that they should not pull their militia off of this duty protecting the county. But they state that you;

"will exert yourselves, remove any discontents explain the great Necessity & Policy of co-operating effectually with our Allies as the best Means to relieve themselves from heavy Expenses & Burthens & bring the War to the happy & favourable Issue we have been long contending for.

We depend much on your Zeal & Activity at this important Crisis & after mentioning that all both Volunteers & Militia will do well to equip themselves as completely as possible especially in Blankets which are not to be had here"[46]

That last sentence harkens back to my point about military recruitment promising something that was not available. On one hand Pennsylvania is going to equip every new recruit with a blanket. But when Pennsylvania tells the militia to report, they tell them you better bring your own stuff because we have nothing to provide you! To reinforce this even more the Supreme Executive Council writes to York and Chester Counties at the same time that if their militia will bring their own guns they will be paid one hard dollar or 60 Continental Dollars. And they reiterate what was told to Cumberland County, to bring your own clothes and blanket because there will be none when they report to General Washington.

Men are in such short supply for General Washington that he orders the German Regiment in Sunbury to withdraw and join his main army. They were in fact part of the Continental Army and not the militia. So, Sunbury will have to fend for itself. General Washington announces that the urgent need for men is to push east on land to New York while the French push from the sea. In the middle is the northern part of the British Army in New York.

With every resource being given to General Washington's main army, his regiment at Fort Pitt on Pennsylvania's frontier is forgotten. Colonel Broadhead with his Continental Regiment at Fort Pitt reports he has had to resort to using military force to gather provisions from the civilian population around him. The civilian populace in turn file charges against him for confiscating food and buildings. It's a very fine line between the fact that they want the military there to protect them, but are not willing to support them in doing so.

As Pennsylvania try's to gather 4000 militiamen to send to General Washington, it places an order with the Board of War for 200 cartridge boxes, 10 drums, 24 drum heads, 20 fifes, 700 muskets, 700 bayonets, 2000 brushes, 3 wire, 9000 flints and 6000 dozen of cartridges. The Board of War reply's that it can't supply it because everything they have is going to the Continental Army - and they are nowhere near filling their needs either.

General Washington has ordered that the 4000 Pennsylvania militiamen are to muster 3 miles outside of Trenton, New Jersey. But General Philemon Dickinson already has his New Jersey Militia there. As the Pennsylvania Militia starts to trickle into Trenton, General Dickinson writes to the Board of War not to send any Pennsylvania Militia there because there is no food nor place for them to stay. He urges the Board of War to keep them out of New Jersey and keep them in

Pennsylvania. General Lacey has been tasked with gathering the Pennsylvania Militia at Trenton and has been scrambling to find a suitable location for them without any luck.

The Supreme Executive Council writes to the Board of War that this is turning into a calamity. General Washington is persistent with ordering the Pennsylvania Militia into New Jersey and New Jersey isn't happy about it. To try and resolve the issue the Supreme Executive Council urges the Board of War on August 17th to let Pennsylvania keep their militia in their own Counties until they are needed. But it turns into a logistic nightmare. Correspondence is slow and most Counties have already sent out their militia and they are on their way to Trenton. When they get there, some have not been paid and are expecting their money when they arrive, which of course there is none. Without money they threaten to go back home. To make matters worse General Lacey says the ones that have been there for a few days are starting to get home sick. He recommends moving camp every few days to keep them busy!

Both Thomas and Richard are called to muster when Cumberland County musters three battalions, the 6th, 7th, and 8th to send to Trenton. Captain Askey's 8th company is again under Lieutenant Colonel James Dunlop and his 6th Battalion and muster August 22nd. Richard Gunsalus is called on August 29th. He reports as a sergeant under Captain Robert Means and his 7th company, 8th Battalion which was led by Lieutenant Colonel Alexander Brown. Serving along with Richard is his 14 year old brother Daniel. Although the minimum age to serve in the militia was 18, family members could serve as substitutes under the age of 18. So Danial was more than likely substituting for his father Jacobus, who was 46 at the time, and within the age to serve in the militia.

Jacobus had a 230-acre farm to manage along with more than likely 2 young children and a wife to tend to. So sending one of his two sons Danial or Samuel, both who were under the age of 18, would have made sense.

Not wanting to send all of his military resources out of the county, Cumberland County Lieutenant Abraham Smith elects to keep one battalion at home to guard the frontier. The other two battalions he elects to send to Trenton. But as the men are mustered they find that there is no ammunition to be had at the armory. Colonel Frederick Watts eventually buys ammunition from a private company, but there is only enough for one battalion, which he gives to the one going to guard the frontier. Without any ammunition it's decided not to send the other two battalions to Trenton either.

Since Cumberland County had the militia mustered they didn't waste the opportunity to use them. The 7th battalion under the command of Lieutenant Colonel James Purdy is operating about 35 miles south of Sunbury on the Juniata River. Captain Askey and the 6th battalion are held in Cumberland County, with the exception of several companies operating in Northumberland assisting their militia. They are also being assisted by Richard's 8th battalion.

According to Captain Means and Private Moses Dickey, members of Richard's Company, they were sent to guard Fort

Potter. Captain Means states in his pension application that he marched his 44 men to Fort Potter and relieved Lieutenant Johnston, who marched to the Buffalo Valley. Private Dickey states in his pension application that in order to cover more ground they were sent out from the fort in groups of 10-12 on scouting duty and stayed at houses away from the fort.

In neighboring Northumberland County 250 to 300 enemy Tories and Indians attack Fort Rice on September 6th, which is about 13 miles from Sunbury. The fort is lightly guarded by about 20 Northumberland Militia. It was previously protected by the German Regiment, but since their departure to assist General Washington the area has been lightly defended. The remnants of the Northumberland Militia and companies from the 6th battalion of Cumberland Militia race to assist the fort. Knowing that the Cumberland Militia under lieutenant Colonel Purdy is operating nearby a message is sent asking for their assistance. Purdy and his 110 men plus 80 volunteers arrive and they are joined by General Potter and a band of Northumberland Militia.

By the time they arrive the enemy had fallen back into the countryside and the fort was still intact, having fended off the enemy. General Potter takes control of the Northumberland Militia and they track the enemy for a while, but they appear to have split up and disappeared into the wilderness.

Meanwhile one of the most traitorous plots in American history was unfolding at the fortification at West Point, New York. After his command of Philadelphia and basic acquittal at court martial, Major General Benedict Arnold was assigned to the fortifications at West Point. While there he had devised a plan to turn the fortification over to the British for a large sum of money, about 20,000 British Pounds. As the commander of the fortification he was deliberately weakening it. He was not ordering repairs and dispersing his men guarding it. All this in preparation to make it easier for the British to attack and take it over.

The plot almost worked, but was discovered on September 23rd when British Major John Andre', who was working with Arnold was captured. He was found with plans for the fortifications at West Point and correspondence from General Arnold. Arnold narrowly escaped capture and fled to British held New York City. General Irvine was ordered immediately to West Point to take over. Upon his arrival he wrote to the Supreme Executive Council on October 1st;

"I was ordered here with my Brigade on the alarm that was occasioned by Arnolds Villainous business.

I made a rapid march and found the place on my arrival in a most miserable situation in every respect – 1800 Militia had been at the Post, but were chiefly Detached on various pretences, those who remained had no post assigned them, nor knew nor had a single order what to do. I have not heard from Head Quarters to day – but I have reason to believe major Andre & Smith must be Hung."[47]

The Smith mentioned in that last sentence was a spy who was working with General Arnold and Major Andre'. He was acquitted of his involvement, Major Andre' was hung, and General Arnold escaped to England.

It's also interesting to note that Richard's cousin, Daniel Gunsalus, had spent most of his time in the military at the fortifications at West Point. He had joined the New York Continental Line in May of 1777 for three years and was assigned to Fort Montgomery, which was located on the Hudson River about 50 miles upriver from New York City. He was then transferred to West Point where he states in his Revolutionary War pension application that he spent his time working on fortifying the facility there. Luckily he was discharged in the spring of 1780 and missed General Arnold who took over in August of that same year.

By the end of October Thomas and Richard are back in Cumberland. And as winter settled over Pennsylvania in December, the 3rd battalion under Lieutenant Colonel Irwin were ordered to march to Trenton. While on their way to Trenton both Irwin and one of his Captains, John Carothers, writes to the Supreme Executive Council about a grievance their men have. The men had mustered and marched out in August of 1779 to assist with the rescue of Fort Freeland, which I had detailed earlier in this book. They said they were gone for a good 15 days and were never paid for their time. So as compensation they state in their letters that they will be taking 3 weeks off of this muster as compensation, and will meet the rest of the 3rd battalion in Trenton 3 weeks later! I just find this pretty bold and have to chuckle at their bravado.

1781

It turns out that all the fuss to gather men for an offensive was much ado about nothing. With all the military preparation there wasn't a major offensive conducted by General Washington on New Jersey and New York City. The French fleet settled down in Newport, Rhode Island and the British maintained their strong hold on the greater New York City area.

Almost all of the major military fighting was now happening in the southern theatre of Virginia and the Carolinas. This by no means meant that the Pennsylvania Militia was out of the fighting. There was still plenty of hostilities on the frontier. And there still was no help coming from the Pennsylvania Line. They had just plain had enough of years of poor living conditions, lack of food, poor enlistment contracts and no pay. In early January in an unprecedented move they mutinied.

The nearly 2,500 Pennsylvania soldiers at their winter quarters in Morris Town took up arms, demanded that their concerns be met, and planned to march on Congress in Philadelphia. They even threatened to desert over to the British. The uprising was forceful and even violent against their commanding officers. It finally ended with Pennsylvania agreeing to let men who enlisted in the early years of the war under poor contracts the option to be discharged or re-enlist. Some re-enlisted with newer and better contracts, but nearly half of the 2,500 men did not. As a gesture of good will to those men who did re-enlist the Supreme Executive Council authorized a onetime payment to every non-commissioned soldier who had enlisted before 1780 a payment of 9 pounds state money above their regular pay.

Because a lot of men decided not to re-enlist there was a huge manpower problem with the Pennsylvania Line, which went through an entire restructuring. Eventually half of the restructured Pennsylvania Line was left in Pennsylvania and the other half were sent to fight in the southern theatre of operations. One regiment that stayed in Pennsylvania was the 4th Regiment under Lieutenant Colonel William Butler, which was sent to Carlisle in Cumberland County. The County now had a full time military presence - well in theory anyway, the 4th Regiment was seriously undermanned.

As the Continental Army started to win more battles, they were subsequently gathering more prisoners. Pennsylvania became really concerned with the German Hessian prisoners of war that were being moved around. Initially they were supposed to be located in the southern states. But with the fighting there, it's decided to move them back north and Congress wants to hold them in Pennsylvania. But Pennsylvania says that's a bad idea because of the heavy German population there. They are concerned that the Hessians will speak to the Pennsylvania Germans in their native language and get them siding with the British cause. Pennsylvania has already had enough with the Pennsylvania Line mutinying, they don't want the German citizens doing the same.

You may be asking yourself how these German prisoners could be mixing with the Pennsylvania public and causing trouble. Shouldn't they be in prisoner of war camps? Well back during this era of warfare there was a gentlemen's code of conduct in dealing with commissioned officer prisoners. They were considered to be honorable men of their word. They would take an oath to be paroled and could live freely within a certain area and amongst the population, but on their gentleman's agreement that they would not escape. Non-commissioned officers and privates were not given this latitude and confined under guard. So, you could have influential German officers freely mingling with the native population.

Congress either didn't care about Pennsylvania's concerns or they had no other choice, and orders the 1,188 British prisoners to York and the 1,487 German Hessian prisoners to Lancaster. To add to the insult of not recognizing Pennsylvania's concerns, Congress orders them to call 400 militiamen up to guard the captured non-commissioned officers and privates.

Pennsylvania isn't going to put up with this without venting their frustration and their refusal to be liable for the prisoners. They inform Congress that they don't have the money to call up the militia, feed the prisoners, nor house them. They state;

"We think it very unequal that when there are 13 States in Union all the Prisoners should be brought to one. We have allways endeavoured to comply with Requisitions when in our Power, but we do not see the least Probability of answering present Expectations in their full Extent. – Having already observed to our Delegates in Congress the Danger of adding to the Dissaffection of the Inhabitants, especially from the Influence of the German officers."[48]

The letter goes on to read that the militia is in shambles and can't guard that many prisoners even if they called out a muster. And that if the Germans escape, they will easily make it to New York City where they will fall back in with the British Army. And that Pennsylvania has dually notified Congress of their issues and if anything bad happens it's not the fault of Pennsylvania!

Congress gets the message that Pennsylvania isn't having it, well at least most of it. They order the British officers to Simsbury, Connecticut and the British non-commissioned officers and privates are to stay in Frederick Town, Maryland. The German officers are to stay in Frederick, Virginia and the non-commissioned officers and privates in Winchester, Virginia. British non-commissioned officers and privates from recent battles fought in the south will be sent to Lancaster, Pennsylvania.

As the 4th Pennsylvania Line Regiment arrives in Carlisle their commanding officer writes to the Supreme Executive Council that Cumberland County is a mess. There is basically no recruitment for his regiment going on, and if there was the bounty for enlisting is the lowest in the surrounding counties. So, no one wants to join his regiment. And for what men he does have there isn't any food in the government stores nor money to buy any.

In early April General Potter at Sunbury writes that there is once again unrest there. As with everyone, he is just about out of ammunition, which has all been sent to the Continental Army fighting in the south. He is awaiting Militia from Cumberland County, but they have not arrived yet. Captain Askey and his company are mustered on April 15th and are sent immediately to Frankstown instead of Sunbury. In route they stop at Standing Stone and several men, including Private Hugh McClure, are left to help guard provisions there. I mention McClure because he details this trip with Askey in his pension request.

Operating in the vicinity of Frankstown were a band of Indians who were killing and kidnapping. In early June a group of militia and volunteers headed north out of Bedford to see if they could find the Indians. At the same time a group of local militia and volunteers from Frankstown left heading south to meet up with the men from Bedford. After meeting up about 3 miles from Frankstown, on June the 3rd, the two groups were ambushed by the Indians they were seeking. The Indian ambush totally surprised the men who were said to have not even gotten off a shot. The ambush killing 30 of the militia and volunteers. The remaining men, some wounded, escaped back to Bedford and Frankstown. Captain Askey was not involved in the ambush and had remained guarding the fortification in Frankstown with the rest of the Cumberland Militia, which totaled about 75 men. After the attack Captain Askey and his men, along with Private Hugh McClure marched about 18 miles North West to Fort Anderson. They remained there about two weeks, marched back to Standing Stone, and after things had calmed down marched back home.

The killing of so many soldiers created a panic in the area. The inhabitants were concerned about the size of the Indian force that could have inflicted so much death on the militia. The Bedford County Lieutenant said that the situation was so dangerous that he was moving his family to Maryland, which he says was safer. Cumberland County in response to the killings sends more militia under Colonel Brown to Standing Stone and Captain Means to Penns Valley to reinforce the fortifications there. Ammunition was tight and each man was only given three musket balls.

Pennsylvania's concern with escaping prisoners is almost realized in late May. The commanding officer of the prisoner of war camp in Lancaster, which is holding around 800 prisoners, uncovers an escape plan. The British had planned to rush through the prison gate when it was opened up to allow for firewood to be brought in. They were then planning to meet a sympathizer in town who would lead them to the ammunition magazine where they would gain access to weapons. Luckily the plot was discovered, and the ring leaders jailed in tighter conditions.

And once again the militia guarding the British prisoners were bumping heads with the Continental Army stationed nearby. At the time of the British escape plot a member of the Continental Army Light Dragoons under the command of Colonel Steven Moylan was placed in the jail. It's unclear what he was in jail for, but some of his comrades decided to break him out. As they approached the jail the militia sentries didn't know who they were and ordered them to stop. One of the Dragoons pulled out a pistol and was immediately shot dead by one of the sentries. As he fell to the ground his pistol discharged and wounded another sentry in the leg. The situation kept going downhill from there. The Dragoons threatened the entire local militia who were keeping their comrade in jail - and threatened to kill the militia sentry who had killed their comrade. For his own safety and protection the militia man who killed the Dragoon was placed in the jail. A little later the Dragoons tried to free their comrade again and one of the Dragoons was shot in the arm. This put the entire town in fear of a riot. And I'm sure watching the entire event unfold in front of them kept the British prisoners amused!

Just in the nick of time the Dragoons were called up to fight in the southern campaign and left Lancaster in late June. I say just in the nick of time not only because of the infighting with the militia coming to a head but also because Lancaster gets flooded with prisoners. British prisoners destined for the New England colonies were held up in Lancaster, doubling the number they already had to well over 1,700. To add to the problem, they have brought women and children with them, over 250. This along with an outbreak in sickness, lack of shelter, and no food causes the prison commander to blow his top in a letter to the Supreme Executive Council.

But with all the bad news coming from the eastern part of Pennsylvania there is some good news at Fort Pitt. Colonel Brodhead took about 300 hundred of his men along with volunteers and went on the offensive against the Indians. They attacked and plundered numerous Indian encampments and had a very successful campaign against them. He claimed that his men brought back 80,000 British pounds in plunder.

And the good news isn't limited to Fort Pitt. The Colonies and French forces are having success on land and at sea. They continue to rack up victories which are weakening the British.

By early July the unresolved prisoner of war issue is getting worse in Pennsylvania. Germans are sent to Reading, British to York Town and the remaining non-convention prisoners to Lancaster. Non-conventional prisoners were those captured after the 1777 surrender treaty of Saratoga, New York. So, Pennsylvania ends up right where they started in regards to housing prisoners. And it really is a humanitarian crisis. Not only are they dealing with prisoners of war, but with their families as I mentioned a few paragraphs earlier. And the logistics of feeding, housing, and guarding these people are just not adequate. Even the Supreme Executive Council writes that they are embarrassed with the total lack of logistical support for these prisoners.

And as usual the total lack of supplies is statewide. The Lancaster militia reports to Sunbury to assist them with fighting the Indians in mid-July. And they arrive without guns nor ammunition, because they needed them to guard the prisoners of war in Lancaster. The Northumberland County Lieutenant totally unhappy with the militia reporting, basically unarmed, writes to the Lancaster County Lieutenant;

"twenty six of your County militia, who is to serve their touer of Duty in this County, but in the Manner they have come here they are of no service to us; fourteen of them wants Guns, and no ammunition here to give them, which is Realy hard if they are Obliged to Return without doing any service to their Country."[49]

It really is amazing that the colonies are able to survive and fend off British and Indian attacks. They are at war on multiple fronts with multiple enemies. The general population is arguably really not concerned about fighting the British. The colonial currency, which there are numerous types - individual Colony money, Continental money, British currency, Spanish currency, to name just the most popular. All add to what should be a total breakdown of any functioning economy. But somehow the Colonies keep stumbling forward, gaining ground on their enemies.

In early August there was a rumor that the British may attempt to land in the Chesapeake Bay at Baltimore. This caused great concern as to whether or not they would try and move up into eastern Pennsylvania and rescue their fellow soldiers held prisoner there. Reclaiming that many men would give a huge advantage to the British Army. To help strengthen the militia guarding the prisoners of war and to assist in moving them away from the possible British advancement, the Cumberland Militia was called up, to include Captain Askey's company. They were notified to be on a moment's notice to march.

The events that were unfolding with the British entering the Chesapeake Bay would within the next couple of months decide the fate of the Revolutionary War. The fighting continues on the frontiers, as do the continual supply problems. Colonel Samuel Hunter writes from Fort Augusta in Northumberland County that he had to send the Lancaster County Militia home early because he didn't have any more provisions for them. And Colonel Brodhead at Fort Pitt writes that his men are on the verge of leaving and the men of the 7th Virginia Regiment and Maryland Corps who were there to help him are leaving because of the lack of provisions.

Colonel Brodhead was already left in a tough spot regarding men. An expedition had recently gotten underway into present day West Virginia and Kentucky. It was being conducted with Continental Line troops from Virginia and Pennsylvania and Militia from Westmorland County. The Line troops were under the command of Brigadier General George Clark and the Pennsylvania Militia under Colonel Archibald Lochry. Just after setting off the Line troops got ahead of the militia as they traveled down the Ohio River. At a point in present day Kentucky Colonel Lochry and his militia were ambushed by British and loyal Indians, killing almost 40 of the 100 militia men. The remaining were captured. General Clark upon hearing of the loss of the militiamen abandoned the expedition.

While researching pension applications I came across several men who had survived the ambush and lived long enough to file pension applications. One of these men was Isaac Anderson. He initially joined the 8th Regiment of the Pennsylvania Continental line in 1776 and fought at the Battles of Stillwater and Edge Hill. At Edge Hill he was shot through the head and left for dead in the snow. Eventually the British came across him and realized he was still alive. They took him prisoner and he was held in Philadelphia, which they occupied at the time. When the British evacuated Philadelphia Isaac was simply left in the hospital bed in which he was recovering from the wound to his head. I'm sure he could have walked away from his enlistment at this point, but instead he joined Colonel Brodhead as an express rider until his enlistment expired. You would think that after all of that he would have had enough of the military. But no, he joined the Rangers as a lieutenant, patrolled around Fort Barr and Fort Muncy, and eventually fell in with Colonel Lochry when he was forming up his expedition. When the expedition was ambushed, he was captured and taken to Detroit, where he spent a month as a prisoner. He was then taken to Montreal in November. In May of 1782 he saw an opportunity to escape and scaled the wall of the prison he was being kept in and made his way into the wilderness. Without any supplies and just the clothes on his back he made his way through the harsh wilderness for over 12 days before reaching Vermont. He was found almost dead and nearly naked, but he had survived. He went on to marry, had 9 children, and lived a full life until his death in 1839. Pretty amazing.

When I think about how the Colonist could have kept the British and Indians at bay while dealing with all of their logistical, military and economic problems I think of this man, Isaac Anderson. Countless men like him is what beat the British and Indians and made this country what it is today. Men that kept on going no matter what. They got shot and they got up and kept fighting. Taken prisoner and escaped. On deaths door step and lived. It came down to the individual man. Superior firepower, logistics, nor economy beat these foes. It was the will of the Colonial man.

In early September Pennsylvania calls up everyone they can to assist with the British threat in the Chesapeake Bay. Every County is called to provide Militia and Light Horse. Once again the County Lieutenants reply a little baffled. They all say that mustering the militia is just about impossible for duty outside the County, especially since most have not been paid for their militia service in over two years! But they try to muster what they can. The Light Horse units seem to be in better shape and more agreeable to muster. Cumberland County reports that;

"I have Directed the light horse of this County to be in Readiness for field Service agreeable to order of Council, their number is about fifty their horses and Equipment Tolerable good only some of their Swords the last time I examined them were too little but they engaged to provide themselves with such as would be proper for the Service and I flatter myself that when they are called to the field they will make a respectable appearance and render essential service to the public."[50]

But the rush to arms is short lived. Within a week the British fleet was defeated by the French in the Chesapeake Bay. This led to the collapse of naval support for the British Army in the south, who were now held up in Yorktown, Virginia. By late October, without supplies and support, they surrendered. This enormous defeat all but sealed the fate for the British in the American Colonies. The British only held a grasp in the New York City and frontier areas.

The fighting slowed to a trickle, but there were two major engagements closing out 1781. They happened in October at the Battle of Fort Slongo on the 3rd and Battle of Johnstown on the 25th, both in New York. Both engagements were won by the Colonists. An interesting fact about the Battle of Slongo is that General Washington awarded the first Badge of Military Merit to a wounded soldier there. This award would later become the Purple Heart Medal.

The war with the British on American soil was pretty much over by this point. This however did not stop British Tories and Indian hostilities on the frontiers, especially those of Pennsylvania.

1782

Even though much of the British Army had been defeated entering 1782, the Colonist kept recruiting and calling up the militia as though they were still in a full-fledged war. There were still plenty of British around the New York City area and frontiers. And most importantly, the British had not given up.

And I got a feeling from reading letters of this time period that the senior players didn't want a sense of false or premature victory to cause the war to slip away from them at the end. General Washington mentions staying focused and not giving up as he announces that he is preparing for another vigorous campaign. But my favorite is that of Robert Morris, the famous financier of the time and head of the Office of Finance, in a letter to William Moore the President of Pennsylvania. In his lengthy letter he lays out how our allies are backing out because they can't afford to support us in the war. And we as a country can't afford to pay anyone to help us fight a war. He ends his letter with what I think is the driving sentiment of the country to end the war and move on;

"*Let us at once become independent. Really and truly independent. Independent of our Enemies, of our Friends, of all but the Omnipotent.*"[51]

In mid-March the Lieutenant of Cumberland County receives orders to send a company of militia to help defend Bedford County. Winter is coming to an end and the Indians and British Tories are expected to start up their fighting campaigns. Cumberland County complains that they don't have any ammunition to arm their militia and that Bedford County couldn't provide provisions to keep the last group of militia sent there fed. So, he awaits instructions before sending them out.

And the hostilities have already started in Sunbury, Lower Smithfield, and even in Cumberland County. Sunbury writes that Indian movements are on the rise. Lower Smithfield where Manuel Gunsales was captured reports more people being taken captive. And William Brown of Ardmagh Township in Cumberland County writes that they are the nearest Township to the frontier in their county and they fear for their lives as there is no military presence there.

Tensions escalate after the massacre of over 95 Indians just west of Fort Pitt on March the 8th. Over 150 Pennsylvania Militiamen attacked Indian camps in present day Ohio. These camps were thought to be hostile and causing trouble in the area. But after the attack they realized that there were non-violent Christian Indians among the hostile Indian captives. Instead of taking all the captives back to Fort Pitt, or releasing the non-violent ones, they kill them all. Accounts say that the Christian Indians sang hymns and recited psalms as they were executed. Reading the initial correspondence indicates that even during the attack the militia knew some of the Indians were friendly because they identified themselves as so. And after they were captured a vote was taken by the militia to take the friendly Indian captives to Fort Pitt or kill them. The vote was to kill them. So the militia clearly knew that they were killing peaceful Indians.

What may have set the militia off into a killing frenzy was their claim that there were articles found among the Indians from recently raided towns of Washington County Pennsylvania. And that 16 people from the area had been recently killed in Indian raids.

Lieutenant Richard Gunsalus is called up on May 25th to serve in a company of militia under Captain Robert Samuels. All the men are from the Kishacauquillis Valley of Cumberland County. They are part of several companies mustered to guard the inhabitants of the area around Fort Potter from hostile Tories and Indians. Fort Potter is on the northern frontier of the county near present day State College.

Ensign John Bell in his pension application states that upon reaching Fort Potter the officers decided to split up the men to cover more ground. Captain Samuels, Lieutenant Gunsalus, and Ensign Bell each took a portion of the company. None of their pension applications mention any fighting. And after their 2-month tour of duty they returned home.

Bedford County was shaken in mid-July with a brazen attack on Hanna's Town. Over 100 Indians and men from the British Kings 8th Regiment attacked the town and nearby Fort Miller. The attackers killed and captured over 20 of the towns people and drove off over 100 head of cattle. This shakes the settlers in the area and many flee.

With hostilities on the rise Captain Askey is mustered along with his company on August the 14th. The Supreme Executive Council is so alarmed at the recent events that they write General Washington and urge him to send out Pennsylvania Line troops to fight on the frontier. They want three expeditions carried out against the Indians. One from Fort Pitt into the west, one from Fort Pitt to the north, and one from Northumberland to the north. But in September General Washington declines to send troops to the frontier. First, because his intelligence tells him that the British have recalled their men operating with the Indians on the frontier and have stopped their push to get the Indians to attack the Colonists. Second, he feels that moving Continental Troops into Indian held territory and attacking them will only create more hostilities, because the Indians will feel as if they need to retaliate and go on the offensive themselves. So, Pennsylvania will have to handle the situation on their own. And they are not happy about it. Colonel Samuel Hunter writes from Sunbury on October the 26th;

"I am sorry to inform you that the Savages still continue their cruel Hostilities against the Inhabitants of this County. The 8th Inst. the Enemy Wounded one man at Wyoming, And took Another Prisoner, the 14th they Killed and scalped an old Couple on Chilisquake, (the name of Martin) about one mile and a half from Col. James Murray's, and took three young Woman Prisoners, being all the family that was in the house. This old couple being Man and Wife. I saw Laying Killed and scalped, And was one that Helped Bury them. The 24th Inst. They Killed and scalped Serj. Edward Lee of Captain Robison's Company, and took one Robr^t Caruthers Prisoner, about two miles from Fort Rice.

This is the way we are served by these Perfidious Enemy after all the Assurances that his Excellency, General Washington, Rec^d of the British permitting no more partys of the Savages to be sent out Against the Frontier."[52]

As the year ends the Supreme Executive Council decides that they need to get a clear handle on what's happening on the frontiers. So they come up with a plan to send several trusted men to visit these areas and make an assessment as to what's really going on.

The war was winding down and the last major engagement between American and British forces was in September. The British along with their Indian allies attacked Fort Henry, which was west of Fort Pitt. The fort held and the attackers disbanded.

1783

Fighting in Pennsylvania shifts to largely focusing on the frontier and resolving land issues with other Colonies.

The frontier of Cumberland County was still a very dangerous place to live. In a pension application from David Criswell, who enlisted as an Indian Spy, he details the violence still occurring there on a regular basis. He was out on muster with Captain Thomas Alexander and his 7th Company, 8th Battalion. They were out pursuing a band of Indians who had been attacking settlers. As they tracked the Indians the Indians came upon the house of David Easton, who resided there with his wife and 5 children. When David Criswell arrived he states that David Easton's wife and 2 smallest children were *"inhumanly butchered."*[54] He says they hastened their pursuit, but they never did catch the Indians nor did they find David Easton and his 3 older children.

There were also two major land disputes that had been going on throughout the war.

First, was the area north of Northumberland County. It was known as Wyoming during this time and had been disputed as being owned by both Pennsylvania and Connecticut. It was settled with Colonist from Connecticut and largely protected by their militia. During the war the Pennsylvania Militia were also sent to protect the area, and they felt they had a legal standing to it.

Second, was the County of Washington, south of Fort Pitt. It had been claimed by both Virginia and Pennsylvania. There was a lot of unrest in the county during the war as both states tried to tax and draw men into their individual militia.

Eventually both the Wyoming area and Washington County were included in the state of Pennsylvania.

There was also the issue of Pennsylvania trying to raise revenue to pay for the debts they had acquired during the war. One of the fastest and easiest ways was to call in all the money owed by those who didn't serve their time in the militia. Most Counties refused to try and collect this money during the war because they felt it put too much of a financial strain on people who were already under constant attack by Tories and Indians. But now the state wanted their money and pushed the counties to collect it.

Another issue was the dilemma of the Militia and Line soldiers not being paid during their time of service. This came to a head in June at the military barracks in Lancaster when the unpaid soldiers threaten to mutiny, to storm the bank there, and take the money due them. After a lot of negotiating loyal soldiers to the state were brought in to squelch the uprising. The ring leaders of the uprising feared being arrested and fled. The remaining soldiers were calmed down and returned to their barracks.

One of the positive aspects of the end of the war was the surplus of weapons and ammunition. The frontier areas were still under siege and these items were shipped in mass to areas such as Wyoming, Sunbury, Hannas town, and Fort Pitt.

On September 3rd the American Revolutionary War officially ended with the signing of a peace accord.

Post War

Thomas and Richard went on to live full lives after the war, Thomas living to be 80 and Richard 82. That's pretty good for what they had both gone through over their lives. And considering that the life expectancy for a white male at the time was 38 years old, they did exceptionally well.

Little is known of Thomas after the war. He passed away before the first widespread Revolutionary War pension laws in 1818; the initial pension act in 1776 was only for those being seriously injured. I mention this because pension applications were probably the only detailed documentation on a person's life back then. Not only would it outline a person's military service but it would also mention where they had lived, where they were born, parents, friends, associates, and anything else that could help identify them and their service.

He is mentioned in minutes from the Supreme Executive Council dated the 7th of November 1785. In these minutes he is referred to as Captain Thomas Askey and he is asking for, and was granted, compensation for the hire of a horse to carry militia baggage. So, I assume he stayed active with the militia as a captain at least up until that point. I have read unofficial family history that he did get involved with the Rangers, but I never could verify this.

We do know that he and his wife had a total of 10 children. And that he passed away on the 24th of November 1807 and is buried at the Lick Run Presbyterian Church Cemetery in Centre County.

Richard was lucky enough to live at least until he was able to file a pension through the 1832 pension act. I say lucky, because earlier pensions were destroyed by a fire where they were stored. His pension claim sheds the most light on his life I have found and probably the only written account of him. When asked about proof of his date of birth he states that it was written down in his mother's Bible. He also goes into detail about several events during his military service, where he ended up after the war, and where he moved around up until his application for a pension. He also mentions several close friends who could verify his military service. They must have been acquaintances after his time in the service because their names do not show up on any of the units he served with.

I did however find some irregularities in his personal accounts of his military service, but it was just with the time frame. And I think that is more than understandable for a 70 something year old trying to remember exactly what year something took place 50 years prior. He also mustered on so many different occasions I'm sure it all seemed to run together. And I found that was a common theme in reading through thousands of pension applications. Most applicants couldn't remember officers they served under or the dates they served. But then again, these men were in their 70's trying to remember a snapshot of events 50 years prior.

In 1784 Richard states that he moved from Cumberland County to Mifflin County and settled in an area called Mechanicsville. In 1785 he married Eustacia Ann"Stachee" Lucas. She was the daughter of a fellow Revolutionary War soldier, Benedict Lucas. The couple had 10 children.

After filing his pension application in 1832 there was some back and forth with paperwork, but in March of 1834 he was granted $40 per year as a soldier's pension. Richard drew that pension for 4 years before passing away on the 15th of March 1838. He is buried with his wife in the Sand Hill Cemetery in Centre County.

Centre County is steep in Gunsalus and Askey history. And there is a grand war memorial at the county seat of Bellefonte. Here the names of Thomas and Richard are but a few inches from each other. I think it's a fitting way to remember their selfless efforts in helping create the United States of America.

SOLDIERS OF THE REVOLUTION.

MATTHEW ALLISON	THOMAS ASKEY	ARCHIBALD ALLISON
PHILIP BARNHART	DAVID BARR	LAWRENCE BATHURST
PHILIP BENNER	JOHN BOGGS	ANTHONY BIERLY
JACOB BROWER	JOHN BRADY	DANIEL BOILEAU
PETER BRUNER	JACOB BRUSUS	BENJAMIN CARSON
JAMES COOKE	HENRY DALE	ELIJAH CHAMBERS
DANIEL DAVID	JOSEPH DAVIS	J. PHILIP DeHASS
JOHN DOUGLASS	PHILIP DENNY	JAMES DOUGHERTY
JOHN ELDER	JOHN DUCK	JAMES DUNLOP
HENRY FARBOW	PETER FLOREY	ROBERT FLEMING
LUDWIG FRIEDLY	JOHN FREDERICK	CHRISTIAN GAST
JOHN GOHEEN	JOHN GLENN	JOHN GARRISON
ANDREW GRAHAM	ANDREW GREGG	JOHN GRAYBILL
HENRY GRENINGER	JOHN HALL	RICHARD GUNSALUS

Plaque on the War Memorial at Bellefonte, Pennsylvania

Resources

1. Map of the scene of operations of the French and Indian War published in Harpers Encyclopedia of United States History circa 1905.

2. The papers of Col. Henry Bouquet, Hathitrust.org Volume 13 page 71 & 72

3. Pennsylvania State Archives series 1 volume IV page 117

4. Pennsylvania State Archives series 5 volume 1 page 180

5. Pennsylvania State Archives series 1 volume IV page 175

6. Pennsylvania State Archives series 5 volume I page 346

7. The papers of Col. Henry Bouquet, Hathitrust.org Volume 13 page 56

8 The papers of Col. Henry Bouquet, Hathitrust.org Volume 13 page 86

9. Pennsylvania State Archives series 5 volume I page 334

10. The papers of Col. Henry Bouquet, Hathitrust.org Volume 14 page 3

11. Thomas Paine, "The American Crisis" December 1776

12. Pennsylvania Historical & Museum Commission

13 Library of Congress

14. Pennsylvania State Archives series 1 volume IV page 155

15. Pennsylvania State Archives series 1 volume V page 393

16. Pennsylvania State Archives series 1 volume V page 393

17. Pennsylvania State Archives series 1 volume V page 631

18. Map of the Battle of Brandywine George W. Boynton

19. Pennsylvania State Archives series 1 volume page 635

20. Spencer Bonsall September 1877, Library of Philadelphia

21. Pennsylvania State Archives series 1 volume V pages 645/6

22. Pennsylvania State Archives series 1 volume V page 672

23. Pennsylvania State Archives series 1 volume V page 700

24. Pennsylvania State Archives series 1 volume V1 page 122

25. Pennsylvania State Archives series 1 volume V1 page 85

26. Pennsylvania State Archives series 1 volume VI page 101

27. Pennsylvania State Archives series 1 volume V1 page 255

28. Pennsylvania State Archives series 1 volume VI page 470/1

29. Pennsylvania State Archives series 1 volume VI page 581

30. Pennsylvania State Archives series 1 volume VI page 636

31. Pennsylvania State Archives series 1 volume VI page 729

32. Pennsylvania State Archives series 1 volume VII page 106

33. Pennsylvania State Archives series 1 volume VII page 131

34. Pennsylvania State Archives series 1 volume VII page 370

35. Pennsylvania State Archives series 1 volume VII page 406

36. Pennsylvania State Archives series 1 volume VIII page 109

37. Pennsylvania State Archives series 1 volume VIII page 131

38. National Archives Publication Number M881 Roll 0089

39. Pennsylvania State Archives series 1 volume VIII page 235

40. Pennsylvania State Archives series 1 volume VIII page 228/9

41. Pennsylvania State Archives series 1 volume VIII page 203

42. Pennsylvania State Archives series 1 volume VIII page 268

43. Pennsylvania State Archives series 1 volume VIII page 312

44. Pennsylvania State Archives series 1 volume VIII page 356

45. Pennsylvania State Archives series 1 volume VIII page 422/3

46. Pennsylvania State Archives series 1 volume VIII page 443

47. Pennsylvania State Archives series 1 volume VIII page 578

48. Pennsylvania State Archives series 1 volume IX page 15

49. Pennsylvania State Archives series 1 volume IX page 292

50. Pennsylvania State Archives series 1 volume IX page 400

51. Pennsylvania State Archives series 1 volume IX page 493

52. Pennsylvania State Archives series 1 volume IX page 657

53. Pennsylvania State Archives series 2 volume III page 167

54. The National Archives State of Pennsylvania pension applications of;

        Isaac Anderson
        John Barnwell
        John Bell
        Thomas Birchfield
        William Black
        Thomas Blair
        James Campbell
        David Criswell
        Robert Carr
        Moses Dickey
        Peter Dooey
        Hugh Drennan.
        Leonard Engler
        James Gallaway
        Danial Gunsalus
        Richard Gunsalus
        Saumel Gunsalus
        John Hunter
        William Kelly
        Hugh McClure
        Robert Means
        Richard Morrow

Samuel Quigley
John Siglar
John Tate
John Torrence
Samuel Witherow

About the Author

Ed Semler retired from the United States Coast Guard in December of 2007 with over 25 years of military service in both the United States Army and United States Coast Guard. In the United States Army he was an enlisted man and was honorably discharged as a Specialist Four (E-4). While in the United States Coast Guard he was enlisted, obtaining the rank of Master Chief Petty Officer (E-9), was commissioned as an officer, and retired as a Lieutenant (O-3E).

After his military career Ed dabbled in teaching at a Vocational Technical School and was a self-employed plumber for several years. As a past time he enjoys writing and playing the guitar, bass, piano, bugle and harmonica.

Fully retired he resides in Schulenburg, Texas with his wife Jana, a retired Air Force senior master sergeant. Please feel free to contact him at mkcm378@gmail.com or through his website www.edsemler.com

His other publications are;

"Around The World," a memoir of his 25 years of service as an officer and enlisted man in the U.S. Army and U.S. Coast Guard

"U.S. Coast Guard Cutter Sherman (WHEC-720) Circumnavigation Deployment 2001" which details the *Sherman's* historic circumnavigation of the globe and deployment to the Persian Gulf in 2001

"The Three Gunsallus Brothers" a story about fighting for Pennsylvania during the Civil War

"Sam Houston & Napoleon Bonaparte Meet On The Civil War Battlefield" a true story of the Walker brothers

"Thoughts On Being A Chief Petty Officer" a take on military leadership

www.ingramcontent.com/pod-product-compliance
Lightning Source LLC
Chambersburg PA
CBHW061326040426
42444CB00011B/2800